Lee Ting Hun
Waseem A. Wani
Eddie Tan Ti Tjih

Edible Bird's Nest: An Incredible Salivary Bioproduct from Swiftlets

Lee Ting Hun
Waseem A. Wani
Eddie Tan Ti Tjih

Edible Bird's Nest: An Incredible Salivary Bioproduct from Swiftlets

The Incredibility of Edible Bird's Nest

LAP LAMBERT Academic Publishing

Impressum / Imprint
Bibliografische Information der Deutschen Nationalbibliothek: Die Deutsche Nationalbibliothek verzeichnet diese Publikation in der Deutschen Nationalbibliografie; detaillierte bibliografische Daten sind im Internet über http://dnb.d-nb.de abrufbar.
Alle in diesem Buch genannten Marken und Produktnamen unterliegen warenzeichen-, marken- oder patentrechtlichem Schutz bzw. sind Warenzeichen oder eingetragene Warenzeichen der jeweiligen Inhaber. Die Wiedergabe von Marken, Produktnamen, Gebrauchsnamen, Handelsnamen, Warenbezeichnungen u.s.w. in diesem Werk berechtigt auch ohne besondere Kennzeichnung nicht zu der Annahme, dass solche Namen im Sinne der Warenzeichen- und Markenschutzgesetzgebung als frei zu betrachten wären und daher von jedermann benutzt werden dürften.

Bibliographic information published by the Deutsche Nationalbibliothek: The Deutsche Nationalbibliothek lists this publication in the Deutsche Nationalbibliografie; detailed bibliographic data are available in the Internet at http://dnb.d-nb.de.
Any brand names and product names mentioned in this book are subject to trademark, brand or patent protection and are trademarks or registered trademarks of their respective holders. The use of brand names, product names, common names, trade names, product descriptions etc. even without a particular marking in this work is in no way to be construed to mean that such names may be regarded as unrestricted in respect of trademark and brand protection legislation and could thus be used by anyone.

Coverbild / Cover image: www.ingimage.com

Verlag / Publisher:
LAP LAMBERT Academic Publishing
ist ein Imprint der / is a trademark of
OmniScriptum GmbH & Co. KG
Heinrich-Böcking-Str. 6-8, 66121 Saarbrücken, Deutschland / Germany
Email: info@lap-publishing.com

Herstellung: siehe letzte Seite /
Printed at: see last page
ISBN: 978-3-659-79255-7

Copyright © 2015 OmniScriptum GmbH & Co. KG
Alle Rechte vorbehalten. / All rights reserved. Saarbrücken 2015

Edible Bird's Nest: An Incredible Salivary Bioproduct from Swiftlets.

Lee Ting Hun[1], Waseem A. Wani[1] and Eddie Tan Ti Tjih[2].

[1]Institue of Bioproduct Development, Universiti Teknologi Malaysia, 81310, UTM, Skudai, Johor Bahru, Malaysia.

[2]Food Technology Programme, Faculty of Applied Sciences, Universiti Teknologi MARA, 40450 Shah Alam, Selangor, Malaysia

Preface

Edible bird's nest (EBN) is one of the commonly used health-modulating foods in Chinese populations. In traditional Chinese medicine, EBN is known for its beneficial effects in treating several ailments including consumptive diseases, tuberculosis, asthma, dry coughs, stomach ulcers, gastric troubles and bronchial disorders. Due to food and medicinal values, EBN is quite an expensive animal bioproduct, and has thus created a lucrative industry in Southeast Asia. It is one of the main contributors of gross domestic product in several Southeast Asian countries like Malaysia, Indonesia, Philipines and others.

This book begins with the introduction of swiftlets; the living factories of EBN. A good understanding of swiftlets is very important for the people of Southeast Asia whether they are involved with swiftlets or not. There is a discussion on the successful development of swiftlet houses for swiftlet farming. Our main aim was to share thoughts on the future development of a swiftlet house. Some laboratory-related issues on cleaning of raw nests are discussed. Neat and standardized scientific procedures for the cleaning of nests are described, which will be useful to the target audience including people in the research and development field, EBN ranchers, etc. More importantly, the recent research carried out on food and medicinal values of EBN are covered. This will help researchers and the general people to know the recently explored food and health values of EBN. Emphasis has been put on EBN toxicity and management for its safe and proper use. This section will be useful to the product development units and the commercial consumers of EBN. The current challenges and future directions of research in the context of EBN are discussed in a separate chapter.

Therefore, this book is a powerful tool for the people who intend to enhance their knowledge on the essentiality of EBN. It will become a helpful and updated reference material for researchers who are actively involved in EBN, and newcomers to this field.

Acknowledgement

First and foremost, we wish to acknowledge the support from our Mentor, Prof Ramlan Abdul Aziz and Prof Dr. Mohd Roji Sarmidi. Without their openness and inputs, this work would not have been executed well. We are also grateful to Institute of Bioproduct Development, Universiti Teknologi Malaysia for providing us the venue, equipment and infrastructure to carry out research on EBN. Not forgotten also are the researchers, our co-workers, undergraduates, post-graduates and staff who are directly and indirectly involved in the overall project.

We would like to acknowledge the financial support from the Center of Excellence: Swiftlets Malaysia and also Ministry of Science, Technology and Innovation (MOSTI), as well as the assistance from Prof Datin Paduka and Dr. Aini Ideris of Putra University of Malaysia.

We would also like to record our appreciation to Swiflet Garden, Sdn. Bhd. for making their swiftlet premises available to us for studying the swiflets' behaviour, and to Glyken, Singapore, Pte. Ltd. for providing the EBN market information.

Last but not least, we would like to thank our family members, who have guided and motivated us spiritually, mentally and physically leading to the successful completion of this book.

Contents (Chapters 1-8):	Page No.
1. **Introduction**	5
2. **Swiflets**	12
2.1 Classification and Nomenclature	15
2.2 Habitation and Geographical Range	17
3. **Edible Bird's Nest**	22
3.1 Collection of EBN	25
3.1 Physical and Biochemical Analysis of EBN	26
3.1.1 Physical Analysis	26
3.1.1.1 Moisture	27
3.1.1.2 Fibre	27
3.1.1.3 Ash	27
3.1.1.4 Protein Profile	28
3.1.2 Biochemical Analysis	28
3.1.2.1 Proteins	29
3.1.2.2 Fats	29
3.1.2.3 Carbohydrates	30
3.1.2.4 Minerals and Essential Metal Ions	31
3.1.2.5 Amino Acids	32
3.2 Quality of EBN	33
4. **House Farming of EBN**	40
4.1 Bird Premises or Bird Nest Ranching	42
4.1.1 Physical and Behavioral Needs of Birds	43
4.1.2 Human Treatment and Animal Management	44
4.1.3 Human Safety and Government Rules and Regulation	45
4.2 Economic Factors of a Bird House	47
4.3 Future Developments of a Bird House	48
5. **Bird's Nest Cleaning Process**	50
5.1 Conventional Cleaning Process	51
5.2 Improvements in the Cleaning Processes	53
6. **Food Importance of EBN**	55
7. **Medicinal Importance of EBN**	62
7.1 Anticancer Properties	63
7.2 Antiviral Properties	65
7.3 Proliferation Effects on Human Adipose-derived Stem Cells	66
7.4 Epidermal Growth Factor like Property	67
7.5 Bone Strength Enhancement	68
7.6 Eye Care Properties	69
7.7 Neuroprotective Properties	70
7.8 Anti-oxidant Properties	71
7.9 Miscellaneous Properties	73
8. **Conclusions and Future Directions of Research**	83
Abbreviations	89

Chapter-1: Introduction

EBN is the dried saliva of swiflets commonly found in the Southeast Asian countries, e.g. Malaysia, Indonesia, Philippines, Thailand and Vietnam (**Fig. 1.1**). Swiftlets are tiny insectivorous birds that often keep on catching insects during flight from nearly first light until dark. Amongst the several swiftlet species in Collocalia genus, only the nests produced by Collocalia fuciphaga, Collocalia germanis, Collocalia maxima and Collocalia unicolor are commercially important due to their human consumption. EBN is called differently in different languages e.g. *Yan Wo* in Chinese, *Sarang Walet* in Indonesian and *Enso* in Japanese. The saliva secreted from the sublingual salivary glands of swiftlets during nesting and breeding season is the main material used in the building of the nests. Generally, the sublingual salivary glands of swiftlets increase in weight from 2.5 to 160 mg, and reach their maximum secretory activity during nesting and breeding season [1]. Nests are built using saliva as a cementing material for binding feathers and other vegetation together. It also ensures firm attachment to the vertical walls of inland or seaside caves [2]. The grading of EBNs is measured on account of their dry mass, the time spent by swiftlets in building nest, and finally the fat and protein content of hardened saliva. As an example, the white nests (**Fig. 1.2**) are almost entirely made from saliva [3], while the black ones (**Fig. 1.2**) are composed of almost 45-55% feathers and small dried leaves [2].

Fig. 1.1: Geographical regions of Southeast Asia enclosed within the dashed boundary provide habitat to swiftlets (the living factories of EBN).

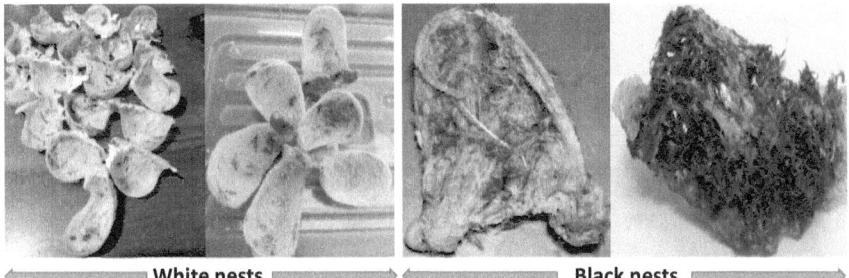

⟵ White nests ⟶ ⟵ Black nests ⟶

Fig. 1.2: A physical overview of white and black EBNs.

The earliest history of EBN trade dates back to almost 1000 years ago in China, during the Tang Dynasty (618-907 A.D.). However, some researchers believe that the EBN trade can only be traced back to 1589 when Ming Dynasty ruled the ancient Chinese Empire [4]. It is said that the Admiral Cheng Ho visited Southeast Asia and was given gift of EBN from Indonesia for the Ming Dynasty's Emperor. This is assumed to be the initial stimulus for opening the trade of these valuable nests. During the ancient Chinese civilization, only the families of the Emperor and his ministers were privileged with the right to consume the highly valued EBN. But, after the end of the monarchy system in China, the general population was introduced to EBN. Owing to the delicious, nutritive and medicinal properties of EBN, its fame rose gradually and steadily among the general public. Nowadays, EBN is reputed as one of the most popular delicacies among the Chinese communities. In addition to the consumption of EBN for health-promoting effects and nutritional components, most of the people consume it as a mere delicious and nutritious food item.

TCM has claimed a long time ago that EBN has highly encouraging effects for the cure and treatment of consumptive diseases, difficult breathing, dry coughs, alleviating asthma, tuberculosis, hemoptysis, asthenia, improving voice, stomach ulcer, relieving gastric troubles and general weakness of bronchial ailments. Besides, EBN is also famous for the proper and healthy nourishment of kidneys, heart, lungs and stomach. Additionally, it helps in raising libido, fortifying the immune system, promoting growth, improving concentration, improving skin complexion, slowing down the aging processes, increasing energy and metabolism, and regulating circulation [5,6]. The regular consumption of EBN has been associated with healthy effects such as high-spirited physical and mental strength, and youthfulness restoration. Proteins form the main ingredients of EBN. They are generally used for the building and repair of body cells and tissues, and driving other metabolic functions. Carbohydrates are another major ingredients of EBN. Sialic acid is one of the major carbohydrates found in EBN, which mediates the distribution and structure of gagliosides in brain. The essential trace elements such as calcium, phosphorus, iron, sodium, potassium, iodine and some essential amino acids are the other main and major ingredients in EBN. In the view of these facts, EBN is a highly nutritive and restorative food with sweet and calm character appropriate for consumption by all age groups of all genders. Recent studies have demonstrated EBN as a potent medicinal substrate with a wide range of medicinal properties including *a*ntiviral, anticancer, *e*ye care, bone strength, neuroprotective and anti-oxidant effects. Besides, EBN has been found to potentiate the p*roliferation of HASCc. In addition, e*pidermal growth factor like activity has been documented for EBN in several experiments.

Presently, the main target market for EBN is the Chinese community all over the world. However, Hong Kong, Mainland China and Taiwan stand out as the top consumers of this product followed by Singapore, U.S.A. and Middle East countries among others. Bird's Nest soup is regarded as an esteemed cuisine by upper class Chinese families and highly appreciated for its health benefits. At diners in expensive Hong Kong restaurants, each bowl of the highest quality bird's nest soup costs almost USD 30-100. There is a stable demand for EBN from the restaurants, however, the demands are at peak during the Chinese New Year period. EBN is usually given as a gift during this period as it symbolizes good health and longevity wishes for the recepient. In addition, it is a symbol of status in society. EBN soup has been a core part of Chinese tradition and culture for hundreds of years now. The increase of wealth in the Asian region, along with a big increase in the price of a bowl of EBN soup has made EBN the 'Caviar of the East' [7].

EBN is a highly expensive animal bioproduct. Approximately one kg of EBN costs USD 6000 in China. The trade of EBN in the global market has been on the rise for decades. It is expected that the demands of EBN will continue to increase irrespective of their low production [8]. Although, EBN has been traditionally prescribed to cure diseases for many generations, the actual characteristics and properties of EBN have not been studied in detail. Besides, there are great commercial opportunities in EBN producing countries. This book was written with the objective to give readers a better understanding of EBN in all aspects *via* a more scientific approach. The underlying objective is to give the readers an insight into the EBN gourmet as a food with medicinal values. Efforts have been made to discuss every aspect of EBN, beginning with the swiftlests; through collection, investigation, house farming, cleaning process and food values to medicinal importance of EBN.

A mind-map overview of the scientific contents discussed in this book is given in **Fig. 1.3**.

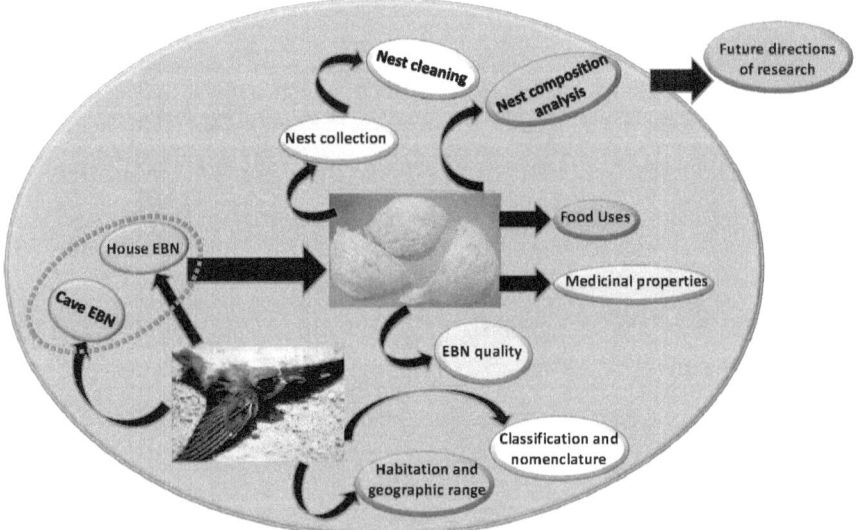

Fig. 1.3: A mind-map representation of the scientific contents discussed in succeeding chapters of this book.

References

1. Medway, L. (1962). The relation between the reproductive cycle, moult and changes in the sublingual salivary glands of the swiftlet Collocalia maxima hume. Proceedings of the Zoological Society of London, 138, 305-315.
2. Kang, N., Hails, C. J., & Sigurdsson, J. B. (1991). Nest construction and egg-laying in edible-nest swiftlets Aerodramus spp. and the implications for harvesting. IBIS, 133, 170-177.
3. Sims, R. W. (1961). The identification of Malaysian species of swiftlets. IBIS, 103a, 205-210.
4. Yeap, T. E. (2002).Edible Bird's Nest Industry in Malaysia. Malaysia: EBN Resource.
5. Hobbs, J. J. (2004). Problems in the harvest of edible bird's nests in Sarawak and Sabah, Malaysian Borneo. Biodiversity and Conservation, 13, 2209-2226.
6. Cranbrook, Earl (1984). Report on the birds' nest industry in the Baram District and at Niah, Sarawak. Sarawak Museum Journal, 33, 143-170.
7. https://sites.google.com/site/ebnresources/history-of-bird-s-nest. Accessed on 02-02-2015.
8. Marcone, M. F. (2005). Characterization of the edible bird's nest the "caviar of the East". Food Research International, 38, 1125-1134.

Chapter-2: Swiflets

Swiftlets are insectivorous birds classified within four genera viz. *Aerodramus*, *Hydrochous*, *Schoutedenapus* and *Collocalia*. They form the Collocaliini tribe within the swift family Apodidae (**Fig. 2.1**) [1]. There are around thirty species of swiftlets mostly confined to southern Asia, south Pacific islands and Northeastern Australia, both within the tropical and subtropical regions.

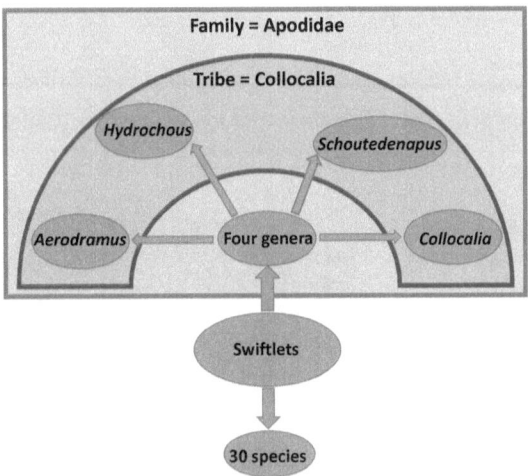

Fig. 2.1: A brief classification of swiftlets.

Generally, swiftlets have narrow wings for fast flight, in addition to a wide gape and small reduced beak surrounded by bristles for catching insects in flight. Nature has bestowed swiftlets with the ability to use a simple and effective echolocation to traverse in complete darkness through the crevasses and channels of caves where they stay at night and breed. Swifts are very aerial species of birds and spend much part of their life spans on their wings. They often fly at high speeds which is assisted by the morphology of their sickle-shaped wings. They have very small feet as is suggested by their family name Apodidae (meaning "feetless") and are therefore, unable to perch. However, their tail feathers are modified which helps them to land on and move on vertical surfaces. The plumage of swiftlets is either dull black or brown, whereas white or gray patches are found in some species, in addition to brighter chestnut-reddish throats in some others [2].

Scientifically, EBN refers to the nest of four species of swiflets, namely *Collocalia fuciphaga* (**Fig. 2.2**), *Collocalia maxima* (**Fig. 2.3**), *Collocalia germanis* (**Fig. 2.4**) and *Collocalia unicolor* (**Fig. 2.5**). *Collocalia fuciphaga* and *Collocalia germanis* produce nests of pure saliva and thus called white nest swiflets while the other two species produce nests with an admixture of feathers and vegetation, and are thus called black nest swiflets [3].

Fig. 2.2: *Collocalia fuciphaga* (Source: http://www.birding.in/birds/Apodiformes/edible-nest_swiftlet.htm).

Fig. 2.3: *Collocalia maxima* (Source: http://www.orientalbirdimages.org/).

Fig. 2.4: *Collocalia germanis* (Source: http://www.orientalbirdimages.org/).

Fig. 2.5: *Collocalia unicolour* (Source: http://www.orientalbirdimages.org/).

2.1 Classification and Nomenclature

There has always been controversy regarding the taxonomy and phylogeny of swiftlets. Morphological and nest characters in addition to the nature of prey, and echolocating ability have been used for the classification of swiftlets [4-6]. All the swiftlets were placed into a single genus "Collocalia" by Gray in 1840 [7]. Gray's classification was used until the discovery of echolocation in swiftlets. Later on, Brooke [8] split the genus Collocalia s.l. into three different genera, *viz.* non-echolocating Collocalia s.s., non-echolocating Hydrochous (comprised of the only giant swiftlet, Hydrochous gigas), and echolocating Aerodramus. However, in subsequent research works, different classification methods were used by different workers. All the above three genera were categorized into

Collocalia s.l. by Chantler and Driessens [9] and Salomonsen [10], whereas Chantler et al. [11] and Sibley and Monroe [12] divided swiftlets into two or three different genera. Most of the authors in the research papers on the investigations of the properties of EBN, did not disclose which swiftlet species were involved in nest building [13,14]. The reason for non-disclosure of the identity of swiflets is the inadequacy of the system of classification as none of the methods is distinctive without enough information to be reliable.

In order to remove the chaos about the taxonomy and phylogeny of swiftlets, several molecular biological studies have been carried out. Lee et al. [15] sequenced cytochrome-b DNA of swiftlets, however, it was only 406 bp portion and thus several questions remained unanswered. Later on, Thomassen et al. [16] sequenced the cytochrome-b gene completely. But, the position of Hydrochous was uncertain due to the high amount of variation in cytochrome-b. Two years later, Thomassen et al. [17] further added two sequences to their original dataset of cytochrome-b sequences for finding the solution to the problem. The two sequences included mitochondrial 12S rRNA (12S) and nuclear non-coding β-fibrinogen intron 7 (Fib 7). The two sequences were believed to evolve more conservatively than cytochrome-b. Besides, the authors also sequenced cytochrome-b and NADH dehydrogenase subunit-2 (ND2) mtDNA of two specimens of H. gigas and found that H. gigas is the sister-group of Aerodramus, which supported monophyly of swiftlets. In the recent past, Lin et al. [18] extracted and sequenced DNA from EBN based on sequence of cytochrome-b gene in mtDNA. The authors successfully identified the genetic source of EBN and its products. It was observed that Aerodramus fuciphagus was the genetic source of Indonesian EBN. Despite the fact that advanced taxonomic methods have been developed, researchers often use the classification described in different times. More often EBNs are classified according to the place of origin [19].

Till the present day, there are no standardized nomenclatures used for naming swiftlets as no consensus has been reached among the researchers on the issue. To avoid confusion, the species of the EBN producing swiftlets in this textbook will use the nomenclature as given in **Table 2.1**.

Table 2.1: Taxonomy of the EBN producing swiflets in context in this textbook.

Superorder	Apomorphae
Order	Apodiformes
Family	Apodidae
Sub family	Apodenae
Tribe	Collocaliini
Genus	Collocalia
Species	*Collocalia fuciphaga*

2.1 Habitation and Geographic Range

Swiflets are insectivorous birds that have a worldwide distribution except for the Arctic and Antarctic regions. Swiflets (Collocaliini tribe) are small swifts that are distributed over areas ranging from the Seychelles islands of the western Indian Ocean through southern continental Asia, Indonesia, Palawan in the Philippines, northern Australia, New Guinea, and the islands in the south-west of Pacific [4,7]. However, EBN-producing swiflets are only found in Southeast Asia [3], in the caves in Adamman and Nicobar Islands, Phuket, Thailand, Malaysia, Indonesia, Singapore and Indochina. The main commercial EBN producers include Indonesia (Sumatra, Java, Kalimantan and the

Lesser Sunda Islands) [8], Malaysia (including Sabah and Sarawak), Thailand, Vietnam [17,20] and Myanmar. Some EBN producing EBN colonies have also been found in Hainan Island in China [16] and Andaman and Nicobar Island in the Indian Ocean [21,22]. However, the produce of EBN in these regions is relatively very insignificant. In Eastern Malaysia, raw EBN is sourced from the caves of Madai and Gomantong in Sabah, the Niah Caves in Miri, the Jade Mountains of Baram and the Dragon Mountains in Tatau and Bintulu [23]. The population of the cave swiflets in Sarawak and Sabah is estimated to be approximately 2 million birds. In Western Malaysia, the main areas of EBN cultivation are Sitiawan, Teluk Intan, Kota Bahru, Kuala Terengganu, Parit Buntar, Bukit Mertajam, Nibong Tebal, Kuantan, Muar, Taiping and many other old townships [24]. However, there is still lack of scientific approaches to precisely estimate the number of birds in Malaysia, especially in Sabah and Sarawak.

References

1. Ma, F., & Liu, D. (2012). Sketch of the edible bird's nest and its important bioactivities. Food Research International, 48, 559-567.
2. Camfield, A. (2004). Apodidae. Animal Diversity Web (online). Accessed on 12-05-2015.
3. Lau, A.S.M., & Melville, S. (1994). International Trade in Swiftlet Nests with Special Reference to Hong Kong. TRAFFIC International, Cambridge.
4. Brooke, R. K. (1970). Taxonomic and evolutionary notes on the subfamilies, tribes, genera and subgenera of the Swifts (Aves: Apodidae). Durban Museum Novitates, 9, 13-24.
5. Mayr, E. (1937). Birds collected during the Whitney South Sea expedition. XXXIII. American Museum Novitates, 915, 1-19.
6. Medway, L. (1966). Field characters as a guide to the specific relations of swiftlets. Biological Journal of the Linnean Society, 177, 151-177.
7. Gray, G. R. (1840). A list of the genera of birds, with an indication of the typical species of each genus. London: Richard and John E Taylor.
8. Brooke, R. K. (1972). Generic limits in old world Apodidae and Hirundinidae. Bulletin of the British Ornithologists' Club, 92, 53-57.
9. Chantler, P., & Driessens, G. (1995). Swifts: A guide to the swifts and tree swifts of the world. Cambridge, MA: Harvard University Press.
10. Salomonsen, F. (1983). Revision of the Melanesian swiftlets (Apodes, Aves) and their conspecific forms in the Indo-Australian and Polynesian region. Det Kongelige Danske Videnskabernes Selskab Biologiske skrifter, 23, 1-112.

11. Chantler, P., Wells, D. R., & Schuchmann, K. L. (1999). Family Apodidae (swifts). In: Del Hoyo, J., Elliott, A., & Sargatal, J. (Eds.). Handbook of the birds of the world. Barnowls to hummingbirds, Vol. 5, Barcelona: Lynx Edicions.

12. Sibley, C. G., & Monroe, B. L. (1993). A supplement to distribution and taxonomy of birds of the world. New Haven, CT: Yale University Press.

13. Pozsgay, V., Jennings, H., & Kasper, D. L. (1987). 4,8-Anhydro-N-acetylneuraminic acid Isolation from edible bird's nest and structure determination. European Journal of Biochemistry, 162, 445-450.

14. Farrar, G. H., Hulenbruck, G., & Karduck, D. (1980). Biochemical and lectin-serological studies on a glycoprotein derived from edible bird's nest mucus. Hoppe-Seyler's Zeitschrift fur Physiologische Chemie, 361, 473-476.

15. Lee, P. M., Dale, H. C., Griffiths, R., & Page, R. D. M. (1996). Does behaviour reflect phylogeny in swiftlets (Aves: Apodidae)? Proceedings of the National Academy of Sciences of the United States of America, 93, 7091-7096.

16. Thomassen, H. A., Wiersema, A. T., Bakker, M. A. G., Knijff, P., Hetebrij, E., & Povel, G. D. E. (2003). A new phylogeny of swiftlets (Aves: Apodidae) based on cytochrome-b DNA. Molecular Phylogenetics and Evolution, 29, 86-93.

17. Thomassen, H. A., den Tex, R. -J., de Bakker, M. A. G., & Povel, G. D. E. (2005). Phylogenetic relationships amongst swifts and swiftlets: A multi locus approach. Molecular Phylogenetics and Evolution, 37, 264–277.

18. Lin, J. R., Zhou, H., Lai, X. P., Hou, Y., Xian, X. M., Chen, J. N., Wang, P. X. Zhou, L., & Dong, Y. (2009). Genetic identification of edible birds' nest based on mitochondrial DNA sequences. Food Research International, 42, 1053-1061.

19. Goh, D. L. M., Chua, K. Y., Chew, F. T., Seow, T. K., Ou, K. L., Yi, F. C., & Lee, B. W. (2001). Immunochemical characterization of edible bird's nest allergens. The Journal of Allergy and Clinical Immunology, 107, 1082-1088.
20. Sibley, C. G., & Monroe, B. L., Jr. (1990). Distribution and taxonomy of birds of the world. Yale University Press, New Haven, Conn.
21. Mardiastuti, A., & Mranata, B. (1996) Biology and Distribution of Indonesian swiftlets with a special reference to Collocalia fuciphaga and Collocalia maxima. In proceedings of the CITES Technical Workshop on Conservation Priorities and Actions on Edible Bird's Nest, 4-7 Nov 1996, Surabaya, Indonesia.
22. Nguyen, Q. P. (1990). Discovery of the Black-nest swiftlet Collocalia maxima Hume in Vietnam and preliminary observation of its biology. Bulletin du Museum National d'Histoire Naturelle Section A Zoologie Biologie et Ecologie Animales 18, 3-12
23. Nguyen, Q. P. (1993). The biological basis of sustainability of harvesting, conservation and development the resource of Edible-nest Swiftlet Collocalia fuciphaga germani Oustalet in Vietnam. PhD. thesis, University of Hanoi.
24. Fan, Z. & He, F. (1996). The status and trade of Edible-nest Swiftlets genus Collocalia in China. In proceedings of the CITES Technical Workshop on Conservation Priorities and Actions on Edible Bird's Nest, 4-7 Nov 1996, Surabaya, Indonesia.

Chapter-3: Edible Bird's Nest

EBN is the hardened salivary material secreted by the male birds of several species of swiftlets during nest building. The birds secrete saliva and use it as a cementing material to bind feathers and vegetation together, and shape them into nests with attachment to the walls of inland or seaside caves [1]. Often the nest weighs 1-2 times the swiftlet's body weight. It only supports the mother and the nestlings. The nest construction is completed in 35 days [2].

Traditionally, EBN is classified both before and after processing. The commonly used classification of EBN is based on the nests' external morphological features such as size, shape, feathering, appearance and color. However, traditional classification of EBN did not include the species of EBN-producing swiftlets thus making the classification of EBN inadequate and confusing. Additionally, further classification is based on the location of collection and the country of origin.

On the basis of shape, EBN has been categorized into several types such as nest cup, nest cake and cracked pieces nest (**Fig. 3.1**). Consideration of the size of nests gives rise to one more class of EBN, fingers grade (unprocessed nest) (**Fig. 3.2**). In terms of the aspects of feathering, EBN has been categorized into premium grade nest, second class grade nest and third class grade nest (**Fig. 3.3**). On the basis of physical appearance, EBN has been classified into grass nest, feather nest, edible bird's nest (**Fig. 3.4**). On the basis of colour, EBN has been classified into white nest, blood nest and Hua Yan (**Fig. 3.5**). On the basis of location, EBN has been classified as Gomantong nest, cave nest, house nest and Sarawak nest (**Fig. 3.6**). On the basis of country of origin, there are several classes *viz.* Malaysia, Thailand and Indonesia nests (**Fig. 3.7**).

Nest cup · Nest cake · Cracked pieces nest

Fig. 3.1: EBN classes on the basis of shape.

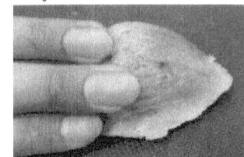

Fingers grade (unprocessed nest)

Fig. 3.2: EBN classes on the basis of size.

Premium grade nest · Second class grade nest · Third class grade nest

Fig. 3.3: EBN classes on the basis of feathering.

Grass nest · Feather nest · Edible bird's nest

Fig. 3.4: EBN classes on the basis of physical appearance.

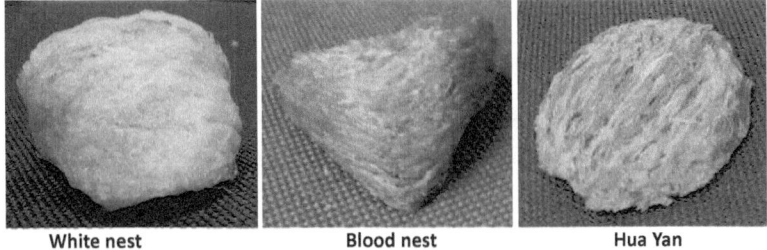

Fig. 3.5: EBN classes on the basis of colour.

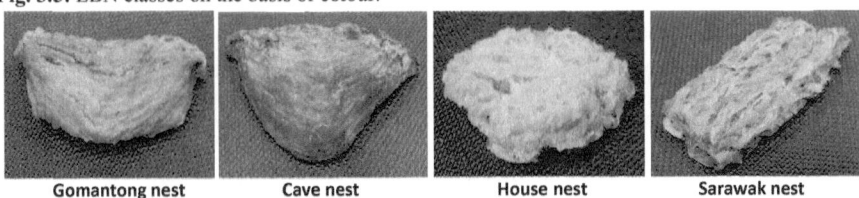

Fig. 3.6: EBN classes on the basis of location.

Fig. 3.7: EBN classes on the basis of country of origin.

Moreover, EBN can also be divided into cave nest and house nest. Cave nests are mainly harvested from natural caves, whereas house nests are made by swiftlets (*Collocalia sp.*) in the attic of countryside houses.

3.1 Collection of Nests

Cave nests are harvested by local people *via* a complex system of management and ownership. The nest harvesting process is often painful and risky [3,4]. The use of nest harvesting techniques depends on several factors such as the cave site, cave height above the ground or water bed, and some other related factors. For the collection of nests in some caves in Kakus, Malaysia, a fishing net is placed across the stream in the cave to catch any fallen nest [5]. For the collection of nests built high on cave walls, temporary frames made from locally collected bamboo or ironwood are used [2]. In low lying caves in Baram, Malaysia, nests are usually collected by hands. The nest collection seasons in Niah and Sarawak, Malaysia, last for 30 to 60 days, while it lasts only two weeks in Baram, Malaysia.

Collection of house nests is a relatively easy task but care needs to be taken of the bird comfort, and also skilled and knowledgeable workers are generally involved.

3.2 Physical and Biochemical Analysis of EBN

Proximate and mineral analysis are some of the predominant methods of analysis for nutritional testing of foods. Proximate analysis represents the crude proteins, fats and fibres, moisture, ash and carbohydrate contents. Many of the food analysis methods in use today are procedures based on a system initially introduced almost 100 years ago [6]. Proximate and mineral analysis is used for the analysis of animal feedstuffs. It involves the estimation of main components of a food using procedures that allow a reasonably rapid and acceptable measurement of various food ingredients without the need for sophisticated equipment or chemicals. Usually the analysed components are crude proteins, fats and fibres, moisture, ash and carbohydrates [6].

3.2.1 Physical Analysis

The physical analysis involves the determination of the contents of moisture, fibres and ash, and the protein profile of EBN.

3.2.1.1 Moisture

The vast majority of methods for the determination of moisture are based on oven drying techniques. Even though these procedures are widely varied, they do not accurately measure water content, and instead measure the volatile matter [7]. The moisture content in EBN was 7.50 % as reported by Marcone [2]. However, publicly available bird's nest moisture content usually varies from 10-50%. It has been observed that some unethical EBN sellers prefer to sell nests with higher moisture content in order to earn more profits. However, EBN with high moisture content will be subjected to the growth of bacteria and fungi, which causes damage and browning of nests. With moisture content controlled below 15%, the EBN shelf-life can be prolonged for a longer time even without refrigeration. It is noteworthy to state that it is impossible to maintain EBN at 0% moisture content in cup shape as it will be fragile to hold its shape.

3.2.1.2 Fibres

Fibres are polysaccharides and lignin, which are resistant to hydrolysis by enzymes in the human alimentary canal. As per this definition, the dietary fibre includes non-starch polysaccharides, resistant starch and lignin [7]. It is important to note that no fibre content was found in EBN by Marcone [2], Saengkrajang et al. [8] and Sarawak Museum Department [9].

3.2.1.3 Ash

Generally, the percentage of ash in food and food products is determined by weight loss after ignition at 525-550 °C. Some types of food matrices call for slightly higher temperatures and numerous sample preparation techniques are recommended for certain types of food products [7]. Processed nests were found to have around 2.5-3.0 % of ash by the Sarawak Museum Department [8]. The reports by Marcone [2] indicated that EBN ash content was around 2.10%. 5.9-7.4% ash content was reported by Saengkrajang et al. [8] in several EBN samples from Thailand. Hamzah and co-workers [10] reported 5.58-13.88% ash content in the EBN samples collected from Malaysia, Indonesia, Thailand and Philippines.

3.2.1.4 Protein Profile

According to Goh et al. [11], the molecular weight of EBN proteins is in the range of 14-97 kDa. This means the size of EBN proteins is relatively small and the separation has to be carried out in a small pore medium. Basically, the pores of agarose gel are relatively large compared to polyacrylamide gels [12]. In addition, many problems have been encountered with the use of agarose for gel electrophoresis. Agarose contains charged groups, principally sulphate and carboxylic acid groups. These groups interact with charged groups on the ionized macromolecules, especially proteins, and hinder their electrophoretic migration. Furthermore, the presence of anionic groups on the support medium leads to an electro-osmotic effect, which alters the electrophoretic mobilities of migrating sample molecules [13]. So agarose gel is less suitable for the protein profiling of EBN.

Some of the reports by Goh et al. [14,11] demonstrated the use of SDS polyacrylamide electrophoresis to identify IgE in EBN, which induces anaphylaxis. The authors also immunochemically characterized the EBN allergen. The protein profiles of commercially available and fresh EBN samples from Sarawak were different. The fresh unprocessed EBNs had more and distinct protein bands. These findings were an indication of the fact that commercial processing may have reduced the amount of intact protein originally present in the fresh nests [11].

3.1.2 Biochemical Analysis

The physical analysis involves the determination of the contents of proteins, fats, carbohydrates, minerals and amino acids.

3.1.2.1 Proteins

Different protein contents of EBN have been reported. Ou et al. [15] characterized the major allergens in EBN using the combined technologies of 2-DE, immunochemistry, N-terminal protein sequencing and MS. The immuno-staining of the Western blots of the EBN 2-DE separated proteins with the sera from allergic patients indicated the presence of a major allergen of 66 kDa. Initial searches of the MALDI-TOF-MS tryptic peptide masses of the allergen in the SWISS-PROT and NCBI non-redundant databases revealed that EBN protein was novel. The average crude protein content in EBN has been reported by Su et al. [16] as 53.26 %, Marcone [2] as 62-63%, Kathan and Weeks [17] as 32.3 %, Saengkrajang et al. [8] as 61.0-66.9%, Hamzah et al. [10] as 59.8-65.4%, Norhayati et al. [18] as 57.9-65.2% and more than 75-85.6% by Sarawak Museum Department [9]. Several studies have been carried out in order to define the precise roles of oligosaccharide chains in the functionality of glycoprotein. Several glycoprotein functions have been identified but many are still under investigation. As EBN consists of mucin types of glycoproteins [2], it can serve as lubricant and protective agent.

3.1.2.2 Fats

Fats are considered a subclass of lipids. The US FDA food labelling regulations define fats as the sum of fatty acids expressed as triglyceride equivalents for nutrition labelling purposes. Saturated fats are fatty acids without double bonds [7]. EBN has very little fat content. The average fat content value reported by Marcone [2] was 0.14%, while Sarawak Museum Department [9] proclaimed the fat content was between 0.2-0.3% by dry weight. Recently, the studies by Saengkrajang et al. [8] on the EBN samples collected from different regions in Thailand indicated that fats formed 0.4-1.3% of the total composition. The fatty acid analysis by Marcone [2] indicated that the EBN fat consisted of palmitic C16:0, stearic C18:0, linoleic C18:1 and linoleic C18:2 acids. The ratio between different fatty acids was dependent on the types of EBN.

3.1.2.3 Carbohydrates

Carbohydrates are a very important class of compounds that are known to form essential food as well as structural components of living species. Carbohydrates are commonly classified into monosaccharides, oligosaccharides and polysaccharides [7]. Phenol-sulphuric acid reaction for carbohydrate analysis of both white and red EBN was carried out by Marcone [2]. The tests indicated that carbohydrates were the second highest occurring components (27.26%) in the entire nest with some differences. Additionally, it was reported that white EBN had slightly more total carbohydrate content than the red EBN. According to Kathan and Weeks [17], the carbohydrate component in EBN consists of 9% sialic acid, 7.2% galactosamine, 5.3% glucosamine, 16.9% galactose and 0.7% fructose. The sialic acid is believed to be N-acetyl-4-O-acetylneuraminic acid. A new, sialic-acid-derived compound was isolated from the acid hydrolysate of EBN by IEC. Combined use of MS, ^{1}H and ^{13}C NMR spectroscopy established that it is the 4,8-anhydro derivative of N-acetylneuraminic acid, and in solutions it exists in two tautomeric forms [19].

A GC detection method was developed to identify the composition of the oligosaccharide chain within the glycoprotein in EBN. This composition includes D-mannitose, D-galactose, N-acetyl-D-galactosamine, N-acetyl-D-glucosamine and N-acetyl neuraminate, which constituted the oligosaccharide chain. The peak-area ratios in GC spectrum for the five monoses were found to be fixed; therefore, the GC technique developed in this work was conveniently used to determine the various raw EBNs and their products both qualitatively and quantitatively. This method serves as a marker to distinguish between the fake and the genuine EBN rapidly [20].

3.1.2.4 Minerals and Metal Ions

For the studies of metal and mineral content in EBN, several characteristics unique to toxicants and nutrients need consideration. Therefore, a distinction must be made between necessary minimal intake and toxic overexposure in the EBN. There has been no evidence of any metal found in the EBN; studied for metal content [9]. The presence of heavy metals in foods may be either due to agricultural processing or from contamination in the food chain.

Several metals have important biological roles within the human body and thus, are considered essential for good health. Such metals are frequently known as minerals. Fourteen minerals have been proved as essential to human health. These essential minerals include calcium, chromium, copper, fluorine, iodine, iron, magnesium, manganese, molybdenum, phosphorus, potassium, selenium, sodium and zinc. Nevertheless, at threshold concentrations, a number of these essential metals become potentially toxic [21]. Minerals are considered as micronutrients as they are needed in relatively small amounts and belong to two groups, *viz.* the macro or bulk minerals; and the micro or trace minerals. Minerals are inorganic elements, and therefore, are not produced by plants and animals. Some of the minerals work as coenzymes, enabling chemical reactions to occur throughout the body [22].

Processed EBN contains reasonable amounts of calcium, iron, and riboflavin. Marcone [2] also detected sodium (650 ppm), potassium (110 ppm), calcium (1298 ppm), magnesium (330 ppm), phosphorous (40 ppm) and iron (30 ppm) in EBN. White EBN was found to be richer in calcium than red EBN. All the red EBNs tested were found to have typically higher levels of iron.

Some researchers believe that minerals found in EBN are leached from the cave substrate where the nests are built. The presence of natural minerals in cave nests and absence in farmed house nets enables the former to withstand longer hours of cooking, whereas the later disintegrates rapidly when cooked [23]. However, there is little published research on the minerals levels in EBN.

3.1.2.5 Amino Acids

Amino acid analysis of EBN involves the identification and quantification of amino acids contained in a particular sample type. Since EBN is consumed as supplementary food, therefore, it is very important to study the protein quality of EBN, which is determined by the amino acid composition. Su et al. [16] developed a capillary electrophoretic method for the determination of amino acid profile of EBN. Eighteen types of amino acids were analyzed with identification of seventeen types including aspartic acid, threonine, serine, glutamic acid, glycine, alanine, valine, isoleucine, leucine, tyrosine, phenylalanine, lysine, proline, histidine, arginine, tryptophan and cystine. Aspartic acid, histidine, proline, serine and glycine were found to be in relatively higher amounts than the other amino acids. It was concluded by the authors that EBN does not contain hydroxyproline and methionine. Besides, the authors did not carry out the composition identification of glutamine and asparagine. Kathan and Weeks [17] reached a similar conclusion stating that EBN is rich in amino acids. Seventeen types of amino acids were identified in EBN, namely, aspartic acid, threonine, serine, glutamic acid, proline, glycine, alanine, valine, methionine, isoleucine, leucine, tyrosine, phenylalanine, lysine, histidine, arginine and cystine. Besides, serine, proline, glutamic acid, threonine and aspartic acid were found in relatively higher amounts. However, the identifications of tryptophan, glutamine and asparagine were not carried out. Newman [24] after some studies on the amino acid composition of EBN documented that EBN is deficient in three essential amino acids, namely lysine, methionine and tryptophan. However, Newman's claims were proved incorrect later on. Marcone [2] identified and quantitated seventeen types of amino acids including aspartic acid, asparagine, threonine, serine, glutamic acid, glutamine, glycine, alanine, valine, methionine, isoleucine, leucine, tyrosine, phenylalanine, lysine, histidine and arginine in EBN. The contents of serine, valine, isoleucine and tyrosine were found to be fairly higher compared to others. However, in his work, the analysis of proline, tryptophan and cystine was not carried out. Furthermore, amino acid analysis revealed that white EBN protein was substantially rich in two specific aromatic amino acids *viz.* phenylalanine and tyrosine.

3.3 Quality of EBN

Due to the fact that cave nests are made by free and naturally-living swiftlets, some people considered them to be more valuable than house nests, and, hence, fetch a higher price. However, in reality, cave nests generally contain more foreign materials and feathers than house nests. Cave nest are exposed to the risk contamination by heavy metals from the external environment. The texture of house nests is usually smoother as compared to that of cave nests. Besides, the house nests have less feathers and other contaminants in comparison to the cave nests.

There are various types of EBN products in the market. The current quality of EBN is not verifiable and its quality claims are totally dependent on the personal validation and rationale made by the manufacturers. Generally, EBN with feathers incorporated in the nest-cup is given lower grades. Good quality nests are distinguished by a comparatively large proportion of nest-cement with few feathers in the nest-cup. After personal observation of EBN trade practices in Malaysia, the colour of EBN is frequently used to grade EBN. Red coloured EBN, also known as red nest or red blood nest is thought to be of higher quality and thus, sells at a higher price. Unfortunately, due to this reason, the white EBNs have been treated with red pigments which are either partially or wholly water-soluble so as to give the false appearance of the red blood nest and hence command a higher price from consumers.

The ancient Chinese communities were traditionally of the belief that the red coloured EBNs were enriched by the blood of swiflets. It is due to this reason such nests are called "Blood Nests" and were considered to be more nutritious. However, it is also possible the area where the nest was built absorbed iron from its environment. Marcone [2] was of the conception that the red tarracota colour of the blood nest is very similar to the colour of the purified ovotrasferrin in its iron complexed state whereas the white coloured nest is similar in colour to ovotransferrin in its iron non-complexed form. Further, X-ray microanalysis reports revealed that the blood nests contained relatively higher levels of iron compared to the white EBNs. Thus, it is likely that the red coloured EBNs are produced via the oxidization of iron in EBNs. Mende [25] suggested that the colour of the EBNs may be due to the nest positions. Besides the above mentioned reasons, it is thought that the colour of nests is affected by the food consumed by swiftlets, which makes the saliva become red naturally. Nonetheless, at this moment, the true reason and factors affecting the colour of the nests are still a mystery.

The quality of the nest has to be monitored from the moment the nests are cultured, sorted, processed and packaged. Judgment regarding the quality of nets merely through the appearance and location of the final nest product will surely be misleading for consumers. In order to produce better quality nests, the nest building or cave environment has to be very clean. Clean environment is very important due to EBN's strong absorption capacities. The nests are often contaminated by lizards, cockroaches, chick's corpse (**Fig. 3.8**), bird's stool and heavy metals, man-made contamination such as pesticides, sodium alginate, starch re-shaping, paints and colours (**Fig. 3.8**), dust, etc. and therefore, a lot of care needs to be taken.

Recently, there have been many reports of fake or adulterated EBN made by adding several materials including fish skin, mushroom, algae, karaya gum, red sea weed, *Tremella* fungus, pork skin, natural plant gum, jelly fungus and egg white [2,26]. These materials are routinely incorporated during commercial processing prior to final sale. Generally, the weight of the nests is increased from 10-30 % to earn extra profits. This a strong offense as it leads to several toxicities in EBN consumers. The presence of nitrites in Malaysian EBN is another issue of concern regarding the safe consumption of nests. The export of EBNs from Malaysia was banned by China on account of the detection of high levels of nitrates in the exported EBNs. It is being estimated that the nitrite and nitrate contaminations in EBNs were contributed by the fermentation process of bird soil and guano [27]. Besides, the contribution of natural environmental resources such as atmosphere, water, and soil was also thought responsible for the contamination of nests [28]. Nitrites are very harmful to human beings as they react with secondary amines and amides in the alimentary canal to form carcinogenic N-nitrosamines [29-31]. Hence, it is a great concern to authenticate the purity of EBN and regulate the law to inspect EBN sold in the market to combat adulteration.

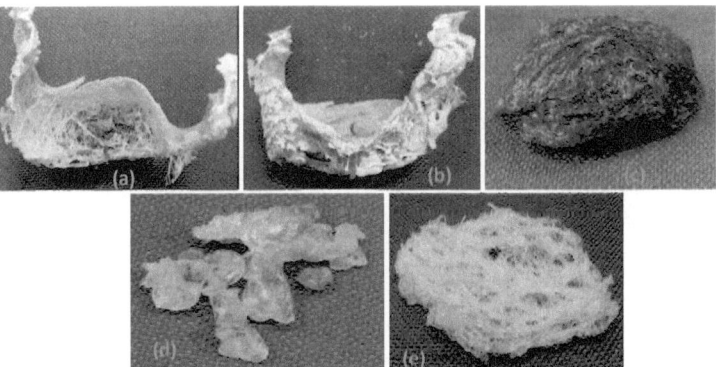

Fig. 3.8: Contamination and adulteration of EBN: **(a)** Nest with chick's corpse, **(b)** Painted nest, **(c)** Nest coloured with red, **(d)** Sodium alginate treated rock nest and **(e)** Nest re-shaped using starch.

References

1. Kang, N., Hails, C. J., & Sigurdsson, J. B. (1991). Nest construction and egg-laying in edible-nest swiftlets Aerodramus spp. and the implications for harvesting. IBIS, 133, 170-177.

2. Marcone, M. F. (2005). Characterization of the edible bird's nest the "caviar of the East". Food Research International, 38, 1125-1134.

3. Hobbs, J. J. (2004). Problems in the harvest of edible bird's nests in Sarawak and Sabah, Malaysian Borneo. Biodiversity and Conservation, 13, 2209-2226.

4. Cranbrook, Earl (1984). Report on the birds' nest industry in the Baram District and at Niah, Sarawak. Sarawak Museum Journal, 33, 143-170.

5. Francis, C. M. (1987). The management of edible bird's nest caves in Sabah. Wildlife Section, Sabah: Sabah Forest Department.

6. James, C. S. Analytical chemistry of foods. Chapman and Hall, 1995.

7. Sullivan DM, Carpenter DE 1993. Methods of analysis for nutrition labelling. AOAC International, Arlington, VA, USA.

8. Saengkrajang, W., Matan, N., & Matan, N. (2013). Nutritional composition of the farmed edible bird's nest (Collocalia fuciphaga) in Thailand. Journal of Food Composition and Analysis, 31, 41-45.

9. Sarawak Museum Department. Bird's Nest Expensive Saliva. Sarawak (Malaysia): Exhibition Brochure, 2004.

10. Hamzah, Z., Ibrahim, N. H., Sarojini, J., Hussin, K., Hashim, O., Lee, B. B. (2013). Nutritional properties of edible bird nest. Journal of Asian Scientific Research, 3, 600-607.

11. Goh, D. L. M., Chua, K. Y., Chew, F. T., Seow, T. K., Ou, K. L., Yi, F. C. & Lee, B. W. (2001). Immunochemical characterisation of edible bird's nest allergens. Journal of Allergy and Clinical Immunology, 107, 1082-1088.

12. Andrews, P. (1965). The gel-filtration behaviour of proteins related to their molecular weights over a wide range. Biochemical Journal, 96, 595-606.

13. Melvin, M. Electrophoresis: analytical chemistry by open learning, London, John Wiley, 1987.

14. Goh, D.L.M., Chew, F.T., Chua, K.Y. & Chay, O.M. (2000). Edible "bird's nest"-induced anaphylaxis: an under-recognized entity? The Journal of Pediatric, 137, 277-279.

15. Ou, K. L., Seow, T. K., Liang, R. C. M. Y., Lee, B. W., Goh, D. L. M., Chua, K. Y., Chung, M. C. M. (2001). Identification of a serine protease inhibitor homologue in Bird's Nest by an integrated proteomics approach", Electrophoresis, 22, 3589-3595.

16. Su, C. S., Yu, P. C., Liu, C. H., Shiau, H. W., Lee, S. C., Chou, S. S. (1998). Application of Capillary for Identification of the Authenticity of Bird's Nest. Journal of Food and Drug Analysis, 6, 455-464.

17. Kathan, R. H., & Weeks, D. I. (1969). Structure studies of Collocalia mucoid I. Carbohydrate and amino acid composition. Archives of Biochemistry and Biophysics, 134, 572-576.

18. Norhayati, M. K., Azman, O., & Wan Nazaimoon, W. M. (2010). Preliminary Study of the Nutritional Content of Malaysian Edible Bird's Nest. Malaysian Journal of Nutrition, 16, 389-396.

19. Pozsgay, V., Jennings, H., & Kasper, D. L. (1987). 4,8-anhydro-N-acetylneuraminic acid. Isolation from edible bird's nest and structure determination. European journal of Biochemistry, 162, 445-450.

20. Yu-Qin, Y., Liang, X., Hua, W., Hui-Xing, Z., Xin-Fang, Z., & Bu-Sen, L. (2000). Determination of edible bird's nest and its products by gas chromatography. Journal of Chromatographic Science, 38, 27-32.

21. Donkin, A. J., Dowler, E. A., Stevenson, S. J., & Turner, S. A. (2000). Mapping access to food in a deprived area: the development of price and availability indices. Public Health Nutrition, 3, 31-38.

22. Lieberman, S., & Bruning, N. Real Vitamin and Mineral Book. Diane Pub. Co., 1997.

23. Leh, C. M. U. (2001). A guide to birds' nest caves and birds' nests of Sarawak, 3rd ed., Heng Sing Brothers Press, Sarawak, pp. 1-20.

24. Newman, J. M. (1995). Usual Ingredients That Some Call Precious, Others Exotic. 2, 11-13. (http://www.flavorandfortune.com/dataaccess/article.php?ID=78. Accessed on 08-02-2015).

25. Mende, R. D. S. Y. (2000). Kajian identifikasi kandungan senyawa bioaktif berdasarkan komposisi zat gizi sarang burung dari burung walet [tesis]. Bogor: Program Pasca Sarjana, Institut Pertanian Bogor.

26. Su, S. C., Yu, P. C., Liu, C. H., Shiau, H. W., Lee, S. C., & Chou, S. S. (1998). Application of Capillary for Identification of the Authenticity of Bird's Nest. Journal of Food and Drug Analysis, 6, 455-464.

27. Paydar, M., Wong, Y. L., Wong, W. F., Hamdi, O. A. A., Kadir, N. A., & Looi, C. Y. (2013). Prevalence of nitrite and nitrate contents and its effect on edible bird nests' color. Journal of Food Science, 78, T1940-T1947.

28. Shirley, R. L. (1975). Nutritional and physiological effects of nitrates, nitrites, and nitrosamines. Bioscience, 25, 789-794.

29. Wolff, I. A., & Wasserman, A. E. (1972). Nitrates, nitrites, and nitrosamines. Science, 177, 15-19.

30. Butt, S. B., Riaz, M., & Iqbal, M. Z. (2001). Simultaneous determination of nitrite and nitrate by normal phase ion-pair liquid chromatography. Talanta, 55, 789-797.

31. Larsson, S. C., Orsini, N., & Wolk, A. (2006). Processed meat consumption and stomach cancer risk: A meta-analysis. Journal of the National Cancer Institute, 98, 1078-1087.

Chapter-4: House Farming of EBN

Due to the high economic value of EBN, swiftlets are being reared extensively in man-made houses in Indonesia, Malaysia and Thailand to increase the production of EBN. In recent years, the price of EBN increased drastically due to high consumer demand. This has led to EBN producers and the Malaysian government to make efforts for increasing the production of EBN through conservation and house farming. In several areas, the conservation of resources has been aided by the domestication of swiftlet species in bird houses.

House farming of swiflets in Indonesia began in the 1800s, and the industry has considerably developed since then. EBN is cultivated by cross fostering, where the eggs of the white bellied swiflets are replaced with the eggs of EBN producing swiftlets, thus establishing a new population of EBN producing swiftlets. While populations of the cave EBN swiflets have been declining due to over exploitation of their nests, rapidly growing house farming of the EBN swiflets has given them a sigh of relief [1].

The Government of Malaysia under the Ministry of Science, Technology and Innovation (MOSTI, formerly known as Ministry of Science, Technology and Environment) together with the Wild Life Department of Malaysia (PERHILITAN) have recently begun encouraging the general public to venture into the very lucrative EBN industry. Over the years, a lot of conservation workshops and laws have been implemented in EBN producing countries to raise the production of EBN. However, the research on the house-collected EBN is scarce as the previously researched samples have been mostly obtained from the caves in Indonesia [2].

4.1 Bird Premise or Bird Nest Ranching

The most interesting thing about these swiftlets is their nest. They construct nests with glutinous strands of starch-like saliva produced by a pair of large, salivary glands and thereafter mate and breed their young ones. It is the nest material that gives them an upper hand over the rest of the birds in the world. The nest looks like a cupped hand or a crescent shaped bowl about 3-5 inches in diameter. The flat side is stuck to a wall (a cave wall in wild nest, or a wooden base in cultivated/house nest) and other side of the nest is a place for them to perch, and within the hollow of the nest are either their eggs or their chicks (**Fig. 4.1**).

Fig. 4.1: Swiftlets enjoying inside the bird house.

The EBN industry is a multi-million dollar enterprise from production, sorting and cleaning to domestic and international trade. According to a survey in December 2001, approximately 1,000 buildings/houses cultivating EBN in Malaysia were reported. The houses produced not less than 10 tons per year (cultivated white bird nest/house nest) while the world demand is 200 tons per year. The demand is still growing but the supply is only able to fulfil half of the demand, which kept the price ever increasing.

Basically, a bird house is a building with artificially created conditions similar to a swiftlet dwelling cave. It is able to attract swiftlets so that they live and breed in it (**Fig. 4.2**).

In designing a bird house, care should be taken of physical and behavioral needs of birds, human treatment and animal management, human safety, and government rules and regulations.

Fig. 4.2: Bird house is a building with artificially created environment to attract swiftlets; **(a)** Shop lots types of bird's house, **(b)** Inner view of bird's house, **(c)** Swiflets entrance of the bird's house, **(d)** Nests in the bird's house and (e) External view of the "bird's hotel".

4.1.1 Physical and Behavioral Needs of Birds

Most of the bird farmers concentrate only on the bird house construction for attracting birds into the houses. This is by far the most important factor in controlling or affecting the bird population in the house. However, it would be better to give consideration to other factors as well prior to the selection of the bird house location. These factors include the swiftlet's flight path, its population at the specific area, nearest source for food and water, and environmental assets.

The most important factor is whether the selected area has the particular species of birds as part of ecosystem. If the area has any of the birds flying through, it would be good. However, if the selected area is quite far away from the bird's habitat area, it may not be a good choice. Furthermore, the birds can be attracted to the bird house by using a special bird recording sound, and then their population can be estimated in that area as shown in **Fig. 4.3**.

Fig. 4.3: Birds attracted by the special bird sound and then estimated. The enclosed portion within the dashed border contains the population of birds to be estimated.

After a particular area is selected, surrounding factors also need a serious consideration. It is also important to find out if the environmental factors in the area under consideration are not supporting. If there is any factory or industry producing heavy toxic materials, smoke or even noise; it is better to avoid that area. Besides, it is quite beneficent to have the presence of supporting factors such as river, pond, forest reserve and open field nearby the selected area.

4.1.2 Human Treatment and Animal Management

Once the selection of the location is made, it is important to plan a bird house or convert a shop house into bird farming premises (bird house). At this stage, human treatment and animal management concerns are very crucial. The information on building bird houses can be had through farming magazines, newspaper advertisements or bird farming association. Expert services are available who can guide on building or converting the shop lot into bird house. The expert credentials are being emphasized when the potential clients are shown a previously constructed successful bird house producing adequate amount of nests. However, most of the experts' advice on account of their experiences without concrete finding, studies or research. The behaviour of birds is scarcely studied as there are only a few publications on the bird's behaviour. There are no records on how many birds were to return their house if they were to hatch. Also no data are available on how many times these birds fly back to their nest daily. Besides, the factor of importance is the bird's willingness to stay back into the origin house or chose a new house for their nest building. However, people have been successfully attracting birds into a house and harvesting bird nest.

Human beings must always treat birds as their companions without staying too close to them. Birds must be given privacy, secured environment and enough comfort to nest their next generation. Environmental guidelines should be followed while applying pesticides and treating the waste of birds. Besides, a great deal of efforts is often required in maintaining the integrity of the nests.

One of the most serious risks associated with bird's life in the bird house is the theft of their nests. The faulty persons break into the bird house and harvest the nests on behalf of the owners. It is a very cruel and off human act to steal nests from the house, as the thief never considers the life of the chicks and eggs in the nest and just drops them on the floor; making this helpless creature to die slowly and with a great deal of torture. Therefore, the bird house owners must consider the security structure and system to secure their wealth and also their wealth creating partners; the swiftlets.

4.1.3 Human Safety and Government Rules and Regulation

There has never been so much emphasis on human safety during nest cultivation. This has totally been a neglected fact in the pursuit for maximum profit. Although surveillance tests conducted by the Department of Veterinary Services from the year 2000 have shown negative results for Newcastle Disease, Avian Influenza or Bird Flu H5N1, there is still risk of getting these viruses. There is a misconception that these bird do not share the same water source with other birds or animals. In reality, they take water from ponds and rivers by flying low on the water surface and occasionally scoping up a sip of liquid. This is particularly true during hot weather and after a long draught. Therefore, it would be wise to take precautions while making entry and exit from the bird house by putting up safety devices to ensure such viruses and other pesticide do not enter or exit from the bird house. There should be contingency plans in the event of an outbreak to segregate people from the birds. However, till now, there is yet to find any example of this setup in any of the bird houses in Malaysia, Thailand, Indonesia or anywhere else in the world.

The Malaysian government has not given any clear guidelines as to the setting-up of bird houses in Malaysia. Therefore, commercial shop houses have been converted into bird houses with people running restaurants downstairs. Some of the bird houses have residents who live there to safeguard their asset. The establishment of bird houses in a residential area might have associated health risks. There are no guidelines and regulations available whether or not bird farming can be done in the housing area or in the middle of the town centre. Care needs to be taken of the children who are sick and coughing when such premises are built around your area. However, the only guideline published by one of the government agencies has discussed noise pollution and the ugly structure leaving the township, an eyesore.

Swiftlets Eco Park by Bio Research Centre (M) Sdn. Bhd. is the one particular development where specific areas and premises were selected to build and farm swiftlets. It is the first and the only one in Malaysia which the government has approved so far. This is an opportunity for all the potential bird farmers to buy one of these premises and start cultivating bird's nests without worrying about the rules and regulations and other annoying factors.

4.2 Economic Factors of a Bird House

To go into the nest business, choose a good location, build a house, install all the necessary facilities to attract birds and start harvesting. All these processes need an investment ranging from RM 300,000-RM 500,000 depending on the location and size of the building. This investment will generate a good and progressive income over the years, while the owner maintains his full time job. This business can be done by everyone. Suppose by having a harvest of 110 to 130 nests that make up to 1 kg of the market price of RM 4,000-RM 6,000 every month after one or two years. The investment will increase by 30 to 100 % where most of the cases are. That is why this is the business of what everyone is talking about.

The funny part in the whole business is how to grade a raw bird nest. Normally, all the bird farmers sell to a middle man by evaluation through experience and making further negotiations on the price. They evaluate the size, color, impurity (bird's feather) and shape (strait or angle), but have little concern on moisture, integrity, pesticide or heavy metal content. This is extremely unacceptable as bird nest is considered as a tonic and most of the time it is consumed by elderly, youth and expecting mothers.

4.3 Future Development of a Bird House

Future development of bird houses depends upon the government moves. It is certain that this industry will not stop here. It is expected to continue its growth due to the lucrative returns. However, more stringent rules and regulations must be installed to reduce or nullify the degree of hazards brought about by this industry. Also, adding a system of taxation into this multi-million industry will be a positive move. The newly licensed bird houses will be in great demand due to two reasons. One of the reasons is that the shop houses converted into bird houses are off welcoming by the public, and the other is the stress of the fear that government may start demolishing the previously established shop houses sometime in future.

References

1. Sankaran, R. (2001). The Status and Conservation of the Edible-nest Swiftlet (Collocalia fuciphaga) in the Andaman and Nicobar Islands. Biological Conservation, 97, 283-294.
2. Su, C. S., Yu, P. C., Liu, C. H., Shiau, H. W., Lee, S. C., & Chou, S. S. (1998). Application of Capillary for Identification of the Authenticity of Bird's Nest. Journal of Food and Drug Analysis, 6, 455-464.

Chapter-5: Bird's Nest Cleaning Process

Nest collection is followed by cleaning processes. Nest cleaning is a very important and time-consuming process. A great deal of care is taken during the cleaning of nests. Nest cleaning is usually performed by means of the conventional cleaning process described below.

5.1 Conventional Cleaning Process

Generally, the harvested EBNs consist of impurities like sand, feathers, egg shells, etc. The conventional cleaning process involves the immersion of nests in water and allowing them to swell. This enables an easy separation of the large feathers from the nest matrix by using forceps. However, care must be taken regarding the elution of the water soluble nutrients in EBN. For the removal of the remaining small feathers, sometimes vegetable oil is used to float the feathers. However, addition of some bleaching agents like hydrogen peroxide to bleach the EBN has been carried out by some producers. Use of hydrogen peroxide reduces the labour cost considerably, as the process is fast. The main drawback with using hydrogen peroxide is that it decolourises the black feathers (impurity) in addition to bleaching the nest colour to white. Therefore, the nests will appear clean but in reality some feathers still remain in the nest.

The hydrogen peroxide treated nests are claimed as safe and of food grade quality but, no producer has ever tried to assess the hydrogen peroxide content in nests after their processing. There are no clear guide lines in any country for the control of the tolerance limit of hydrogen peroxide content. Thus, the conventional method of nest cleaning suffers over health concerns. Hydrogen peroxide is a strong oxidant and harmful to the human body. Therefore, in order to avoid the hydrogen peroxide processed nests, it would be better for consumers to choose nest cups rather than nest cakes. The effect of bleaching on the physical appearance of EBN can be seen from **Fig. 5.1**.

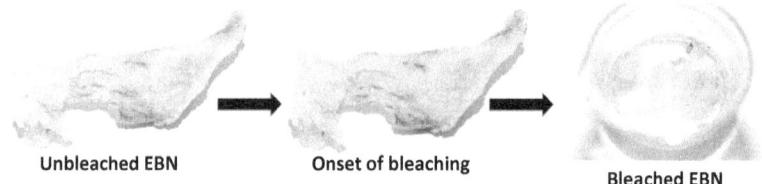

Unbleached EBN   **Onset of bleaching**   **Bleached EBN**

Fig. 5.1: Changes in the physical appearance of EBN during and after the bleaching process.

After cleaning process, the separated strands of EBN and broken filaments are arranged to make nest cake by molding. The shape of the mold could be in leaf shape, round or square, and it depends on the producer's preference. This is followed by drying of the nests. The drying involves the use of fan or the nests are simply air dried. Many workers do not have adequate hygienic food processing knowledge, hence, they do not dry the nests quickly. This gives enough time to bacteria and fungi to grow and proliferate on the nests. Even the colour of some nests turns brownish or yellowish, and yet, the sellers call it the natural colour of the nests.

From the above discussion of the conventional cleaning methods and practices involved in the EBN cleaning industry, it can be seen that the conventional cleaning process of EBN has the following drawbacks.

- Use of bleaching agent, which may be harmful or even destroy some essential nutrients within the treated EBN.
- Additives degrade the nutritive and medicinal values of EBN.
- Preservatives also degrade the quality of EBN as a food and medicine.

- No care is taken of the preservation of nutrients during the treatment processes.
- Generally, tap water is used.
- The process uses the services of untrained workers.
- Little or no regard is given to hygienic practices.
- Nest cakes are produced.
- Generally, there is no quality control protocol being followed.

5.2 Improvement in the Cleaning Processes

Generally, premium grade and quality raw bird's nests are chosen from controlled environment to avoid the presence of any heavy metals or other biological contaminants. Skilled and well trained workers under the supervision of an expert are allowed to separate the feathers and dirt. High grade reverse osmosis water is used to ensure the cleanliness, and preserve the natural flavour and nutrients of the nest. Processed bird's nests are dried in oven (no fan drying) until the right moisture content is achieved. Besides, implementation of the stringent quality control protocols for the end products will allow better grade of bird's nest. With the advancement of knowledge in the food processing technology, some of the requirements which are stated below should be implement in the bird's nest cleaning industry. This will surely allow a better and healthy growth for the industry and increase the quality of the nests, since the public has become more concerned and aware of the food they consume.

- Use of bleaching agent should not be encouraged.
- Additives should be avoided.
- No preservatives should be used.
- Care should be taken of the preservation of nutrients.
- R.O. water should be used for the cleaning processes.

- Highly skilled workers should be involved with the cleaning processes.
- Hygienic practices should be encouraged (**Fig. 5.2**).
- Whole nest cups should be developed.
- Quality Control is a must.

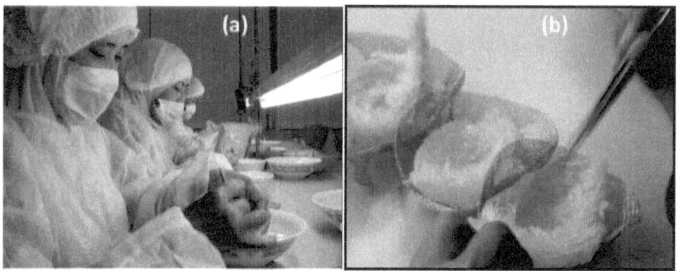

Fig. 5.2: Implementation of hygienic practices in the cleaning of EBN; (**a**) Neat cleaning conditions with skilled workers, and (**b**) De-feathering and moulding by using sterilized forceps and mould.

Chapter-6: Food Importance of EBN

EBN is one of the most popular delicacies among the Chinese communities due its valuable nutritional components and health-promoting effects. Most of the people in the Southeast Asian region consume EBN as a delicious and nutritious food. Nature has gifted EBN with high value nutritional components (**Tables 6.1-6.4**). The findings regarding the compositional ingredients of EBN in these tables fully justify it as a prized delicacy and effective medicine.

Table 6.1: Total Composition of EBN [1-3].

S. No.	Component	Content (Percent proximate analysis)
1	Moisture	7.5-12.9
2	Ash	2.1-7.3
3	Fat	0.14-1.28
4	Proteins	42-63
5	Carbohydrates	10.63-27.26
6	Total nitrogen	25.62-27.26

Table 6.2: Amino acid composition of EBN [1,2].

S. No.	Amino acid	Content (Molar percent)
1	Aspartic acid + asparagine	2.8-10.0
2	Threonine	2.7-5.3
3	Serine	2.8-15.9
4	Glutamic acid + glutamine	2.9-7.0
5	Glycine	1.2-5.9
6	Alanine	0.6-4.7
7	Valine	1.9-11.1
8	Methionine	0-0.8
9	Isoleucine	1.2-10.7
10	Leucine	2.6-3.8
11	Tyrosine	2.0-10.1
12	Phenylalanine	1.8-6.8
13	Lysine	1.4-3.5
14	Histidine	1.0-3.3
15	Arginine	1.4-6.1
16	Tryptophan	0.02-0.08
17	Cysteine	2.44
18	Proline	2.0-3.5

Table 6.3: Fatty acid triacyl glycerol and vitamin composition of EBN [1,2].

S. No.	Fatty Acids (%analysis)/Triacyl glycerols (%analysis)/Vitamins (mg/100g)	Content
1	Palmitric C16:0	23-26
2	Steric C18:0	26-29
3	Linoleic C18:1	22
4	Linolenic C18:2	26
5	PPO	14-16
6	OOL	13-15
7	PLn Ln	18-19
8	Monoglycerides	27-31
9	Diglycerides	21-26
10	Vitamin A	0.0771-0.912
11	Vitamin D	0.15-3.19
12	Vitamin C	0.12–29.30

Table 6.4: Mineral composition of EBN [1-3].

S. No.	Element	Content (ppm)
1	Sodium	330-20,554
2	Potassium	110-2645
3	Calcium	798-14,850
4	Magnesium	330-2980
5	Phosphorus	40-1080
6	Iron	30-1860
7	Sulphur	6244-8840
12	Manganese	3.58-122.10
13	Zinc	19.95-72.40
14	Copper	4.68-110.65
15	Molybdenum	0-0.94
16	Cobalt	0-0.63
19	Nickel	0-0.47
20	Vanadium	0.03-2.84
21	Chromium	0-7.45

Regular consumption of EBN as a tonic ensures high-spirited physical and mental strength in addition to youthfulness restoration. Besides, it improves skin complexion and retards the aging processes. Proteins form a considerable fraction of the composition of EBN. These help in the building and repair of body cells and tissues, and driving other metabolic functions. Carbohydrates form another major portion of the composition of EBN with sialic acid as one of the major carbohydrates. Sialic acid has been known for mediating the distribution and structure of gagliosides in brain. Exogenous supplies of sialic acid enhance the neurological and intellectual processes in infants. Thus, sialic acid in EBN enhances the intellectual and mental development, however, its nutritional and biological mechanisms in human body are not fully known yet.

Essential trace elements such as phosphorus, calcium, sodium, iron, potassium, iodine, and essential amino acids are some main and major ingredients in EBN. In response to these facts, EBN acts as a highly nutritive and restorative food with sweet and calm character, and can be consumed by all age groups of all genders. The frequent consumption of EBN by ladies makes their complexion fairer, and gives them radiant look. Additionally, during the periods of pregnancy in women, its consumption improves immunity of fetus and helps in easy recovery from debilitation after child birth. The health effects of EBN in elderly people include strengthening of lungs and kidneys, phlegm clearance, fortifying of spleen and enhancing appetite. EBN helps to enhance immunity in children. It is quite noteworthy that EBN extracts contain anti-common flu properties and, therefore, EBN may be regarded as a safe and natural source of proteins for the prevention of influenza [4].

Due to the growing fame of EBN as a health modulating and delicious food, researchers have always found ways to develop food products based on EBN and its contents. Li et al. [5] prepared a health food package comprised of independently packaged nutrient powder and nutrient soup. The nutrient powder was composed of pine pollen (0.01-0.5), royal jelly (0.1-5), taurine (0.1-5), milk calcium (0.5-5), α-lactalbumin (0.5-4), oligopeptides (0.5-4), xanthophyll (0.1-0.5), radix puerariae extracts (0.5-5) and poria cocos extracts (0.5-5) by weight parts. Besides, the nutrient soup was composed of rehydrated EBN (0.1-2) and decoction solution of Ziziphus jujuba (70-75) by weight parts. The health food package served as an immune booster to the middle-aged and the elderly. Lian and co-workers [6] prepared a healthy beverage by combining water extract of EBN (200-300), taurine (0.5-1.0), xylitol (50-60), xylooligosaccharide (65-80) and honey (15-30) by weight parts. However, xylitol can be substituted by maltitol. The beverage served as a nutrient rich food item with the merits of good taste and easy storage. Kim et al. [7] prepared a functional beverage with natural healing, skin-improving, anti-aging and immunopotentiation effects. The beverage was prepared by mixing EBN or its extract, red ginseng extract, in addition to more than one of sweetening agent, organic acid, organic acid metal salt, flavour and fruit juice. Besides, the beverage contained D-sorbitol, gellan gum, sodium citrate, calcium lactate, and citric acid. The beverage served as a healthy drink suitable for all age groups.

Some of the famous food products having EBN as the sole or one of the ingredients are shown in **Fig. 6.1: (a-i)**.

Fig. 6.1: Market looks of some of the famous EBN based foods and food products. **(a)** Bird's nest soup, **(b)** Instant bird's nest energy drink , **(c)** Vietnam bird's nest powder, **(d)** Brand new concept EBN powder, **(e)** Bird's nest drink, **(f)** Bird's nest pudding recipe, **(g)** Instant Malaysian cubilose

nourishing tonic, **(h)** Bird's nest granules for supplements and **(i)** Sugar free premium bird's nest extract for consumption.

Overall, EBN is a complete food enriched with a diverse pool of proteins, lipids, amino acids, carbohydrates, minerals and vitamins. This food either as bird's nest soup or in the form of beverages has got exciting health as well as nutritional benefits.

References

1. Marcone, M. F. (2005). Characterization of the edible bird's nest the "caviar of the East". Food Research International, 38, 1125-1134.

2. Lu, Y., Han, D. B., Wang, J. Y., Wang, D. R., He, R. Y., & Han, L. X. (1995). Study on the main ingredients of the three species of edible swift's nest of Yunnan province. Zoological Research, 16, 385-391.

3. Wang, C. C. (1921). The composition of Chinese edible birds' nests and the nature of their proteins. Journal of Biological Chemistry, 49, 429-439.

4. http://www.swiftletecopark.com.my/knowledge_articles_benefits.htm. Accessed on 24-12-2014.

5. Li, A., Liu, C., & Guan, Y. (2012). Health food containing traditional Chinese medicines for the middle-aged and the elderly. Patent No.: CN 102429218.

6. Lian, J., Fan, Q., & Huang, D. (2012). Method for producing beverage containing bird's nest acid. Patent No.: CN 102771852.

7. Kim, W. G., Lee, W. G., Kim, J. S., & Park, G. J. (2010). Functional beverage composition containing EBN and red ginseng extract. Patent No.: KR 2010027265.

Chapter-7: Medicinal Importance of EBN

EBN has been used as medicine for a long time. Literature witnesses the consumption of EBN as "bird's nest soup" by Chinese populations from almost 1200 years ago. TCM recommends EBN as one of the most important bioproducts with health improving effects such as growth promoting, anti-aging, immunity boosting, anti-cancer, anti-aging, dissolving phlegm, immunity-enhancing, alleviating asthma, curing tuberculosis, stomach ulcers and hematemesis, suppressing cough, improving voice, etc. The uniqueness of EBN in TCM is mainly due to its dual nature as it is being used as medicine on one hand and food on the other hand. There has been a significant amount of research on ENB as medicine, and some of the findings are discussed in the following sub-sections.

7.1 Anticancer Properties

Cancer stands as the second most deadly disease after cardiovascular diseases [1-5]. It has created a major public health havoc globally and, therefore, the agents for its treatment have always been in great demands. Natural products as anticancer agents have always been appealing to researchers owing to their promising effects within large safety margins. Presently, several natural product-based anticancer agents are used for treating different cancers. Rashed and Nazaimoon [6] demonstrated the effects of EBN on the proliferation of Caco-2 cells by using the MTT assay. The EBN samples were collected from the Department of Wildlife and National Parks, Kuala Lumpur comprising of two commercial brands and four unprocessed samples. Only 84 and 115% cells were found to proliferate on treatment with the two commercial EBN samples. However, 91, 35 and 47% cell proliferations, respectively were reported on treatment with unprocessed EBN samples from East Coast, North and South Zones. These results suggested the anticancer potential of EBN.

Complementary and alternative medicine (CAM) defines a group of medical and health care systems, practices and products, which are not considered as part of conventional medicine [7]. In Singapore, both cancer patients from western and eastern cultures were exposed to CAM ranging from health supplements to TCM, traditional Malay (Jamu) medicine and traditional Indian (Ayurvedic) medicine. CAM usage is quite popular among cancer patients. Lim et al. [8] demonstrated the use of CAM in paediatric oncology patients in Singapore. Dietary changes, health supplements, herbal tea and EBN were the main therapeutic ingredients of CAM. Shih et al. [9] documented the usage of CAM in Singaporean adult cancer patients. An interviewer-administered questionnaire was completed by 403 adult cancer patients under treatment at the Ambulatory Treatment Unit of National Cancer Centre Singapore. Among all the patients, 46% testified CAM usage including TCM, bird's nest and special diet. 54% of respondents updated their oncologists about CAM usage and interestingly, 66.4% of the oncologists agreed with CAM usage. The effective working of CAM against cancer was felt by majority of the patients. This report indicates the benefits of the consumption of EBN in cancer patients, however, it is very important for health-care professionals to be updated with CAM research and to actively provide appropriate advice and counselling.

Of course, the studies involving the anticancer evaluation of EBN and its extracts have not been carried out over a large range of cancer cells. Only one preliminary study was found in the literature. One of the reasons for the lack of research may be that EBN is a rich food and, therefore, a growth promoter, and it is unlikely to kill cancer cells. However, this may not be true as well. Therefore, it is important to screen EBN over a range of cancer cell lines before any meaningful conclusion can be drawn.

7.2 Antiviral Properties

Viruses are microscopic infectious agents that replicate only inside living cells of other organisms. They infect all life forms including animals, plants, bacteria and archaea [10]. Viruses exhibit several structural and biochemical effects on host cells, the cytopathic effects. Most of the viral infections lead to the lysis of cells, alterations in cell membranes and ultimately the death of host cells [11]. Cold, influenza, chickenpox, cold sores, AIDS, avian influenza and SARS, etc. are some of the most common diseases caused by viral infections.

Influenza is a viral infection caused by influenza virus. Its symptoms include high fever, sore throat, runny nose, muscle pains, headache, coughing and tired feeling. EBN helps to neutralize influenza virus infection in MDCK cells and also causes the inhibition of hemagglutination of human erythrocytes caused by influenza A viruses [12,13]. After hydrolyzation with Pancreatin F, EBN inhibits the infection caused by human, avian, and porcine influenza viruses in a host range-independent manner. However, EBN does not inhibit influenza virus sialidase, and the active inhibiting ingredient of EBN is susceptible to neuraminidase of influenza virus of all strains. The Collocalia mucoid is an established substrate for influenza virus sialidase [14,15], wherein the inhibition can be destroyed by neuraminidase to some extent. Owing to the activities of EBNE against influenza viruses, the presence of a mixture of inhibitory substances in EBN was suggested. It was demonstrated by further studies that N-acetylneuraminic acid, which is the major ingredient in EBN might be responsible for this activity [14]. It was very interesting to note that EBNE showed no side effects such as hemolysis and cytolysis on erythrocytes and MDCK cells even at high 4 mg/ml. Thus, EBN with molecules smaller than 25 kDa after

Pancreatin F treatment will be an effective and safe material as anti-virus [13]. Further studies were carried out in this direction by Yagi and co-workers [16] who demonstrated the N-glycosylation profile of EBN. A tri-antennary N-glycan containing the alpha 2,3-N-acetylneuraminic acid residues was displayed as a major component. The sialylated high-antennary N-glycans were thought responsible for the inhibition of influenza viral infection.

A thorough look into the literature indicated that there are not so many studies that could fully demonstrate the antiviral properties of EBN, and therefore, further studies are needed against different pathogenic viruses to fully explore the antiviral properties of EBN. Further, it would be more advantageous if some studies are carried out wherein EBNEs are synergistically evaluated for antiviral properties with other antiviral agents.

7.3 Proliferation Effects on Human Adipose-derived Stem Cells

Stem cells represent a class of undifferentiated cells with ability of self-renewal, and differentiation into more than one types of cells. Generally, adipose stem cells (ASCs) occur in almost every type of white adipose tissue. The pluripotent ASCs differentiate into most of the mesenchymal cell types including adipocytes, chondrocytes, osteoblasts and myocytes [17-24]. The mesodermal origin of adipose cells makes unlikely their differentiation into neural tissue of ectodermal origin [25]. However, *in vitro* exposure to anti-oxidants makes adipose cells assume a bipolar morphology similar to neuronal cells [26-28]. Stem cells are functionally vital for the repair or regeneration of damaged or diseased tissues. ASCs have been suggested as the best among the mesenchymal stem cells because of sufficient revelations of their pluripotency, proliferating power and low donor morbidity [29]. They are attractive candidates in regenerative medicine because they can be harvested in large numbers with low donor-site morbidity.

Roh et al. [30] reported the proliferation of hADSCs by treatment with EBNEs. EBNE strongly promoted the proliferation of hADSCs *via* the production of IL-6 and VEGF. IL-6 and VEGF production was triggered by the activation of AP-1 and NF-κB. Interestingly, the production of IL-6 and VEGF was promoted by EBNE. The production of IL-6 and VEGF was inhibited by PD98059 (a p44/42 MAPK inhibitor), SB203580 (a p38 MAPK inhibitor) and PDTC (a NF-κB inhibitor), but not SP600125 (a JNK inhibitor). Similarly, EBNE-induced proliferation of hADSCs was also reduced by PD98059, SB203580 and PDTC but not SP600125. This report favoured the fact that EBNE-induced proliferation of hADSCs primarily occurred through augmented expression of IL-6 and VEGF genes, which was mediated by activation of NF-κB and AP-1 through p44/42 MAPK and p38 MAPK.

7.4 Epidermal Growth Factor like Property

Epidermal growth factor (EGF) causes the proliferation, growth and differentiation of cells by binding to its receptor EGFR. Human EGF is a 6045 Da protein containing of 53 amino acid residues with three intramolecular disulfide bonds [31,32]. EGF is known to bind to epidermal growth factor receptor (EGFR) with high affinity on cell surface. This binding interaction stimulates ligand-induced dimerization [33], which in turn activates the intrinsic protein-tyrosine kinase activity of the receptor. The activation of tyrosine kinase activity results in a signal transduction cascade causing several biochemical changes within the cell *viz.* the rise in intracellular calcium levels, increased glycolysis and protein synthesis, and the expression of certain genes including the gene for EGFR. All these cellular changes ultimately lead to DNA synthesis and cell proliferation [34]. It was Kong et al. to demonstrate for the first time that some ingredient is present in EBNE with EGF-like activity [35]. The authors observed that the EGF-like agent in EBN stimulated thymidine incorporation in quiescent culture of 3T3 fibroblasts.

A critical analysis indicates that the EGF-like component of EBN may be responsible for its rejuvenating properties. However, studies are needed to identify the substance, elucidate its structure and explore its possible potential for other biological effects alone; and in EBN as a formulation both *in vitro* and *in vivo*.

7.5 Bone Strength Enhancement

Bones are hard and strong structures in human body forming the skeleton system, and provide enormous support and protection to important organs of the body. Besides, bones are the reservoirs of red and white blood cell production, store minerals in addition to helping in movements and locomotion. It is a well-known fact that strong bones make a strong body and *vice versa*. Matsukawa and co-workers [36] documented the enhancement in bone strength and dermal thickness due to supply of EBNE in the diet in ovariectomized rats. The authors observed that oral administration of EBNE enhanced calcium concentration and therefore, bone strength in the femur of ovariectomized rats was improved. Additionally, dermal thickness also increased by the administration of EBNE. However, EBNE had no effect on the serum estradiol concentration. These results were an indication of the fact that EBNE was effective in the improvement of bone strength and skin anti-aging in postmenopausal women.

Osteoarthritis (OA); a degenerative disease degrades joints including articular cartilage and subchondral bone. This disease is characterized by acute pain, and often causes loss of ability and stiffness. EBNE has been documented to contain some important ingredients reducing the development of osteoarthritis and helping in the regeneration of cartilage. The effect of EBNE on the catabolic and anabolic biochemical activities of the human articular chondrocytes (HACs) isolated from the knee joint of OA patients was described by Chua et al. [37]. The study indicated that 0.50-1.00% of the EBN hot-water extract addition promoted the proliferation of HACs. Besides, the reduction in the expression of catabolic genes such as matrix metalloproteinases (MMP1 & MMP3), Interleukin 1, 6 and 8 (IL-1, IL-6 & IL-8), cyclooxygenase-2 (COX-2) and inducible nitric oxide synthase (iNOS) in cultured HACs was observed due to EBN supplementation. Additionally, prostaglandin E2 (PGE2) production was significantly reduced in HACs. However, type II collagen, Aggrecan and SOX-9 gene expressions in addition to sGAG production was increased as revealed by anabolic activity assessment. This report revealed the *in vitro* chondro-protection potential of EBNE on human articular chondrocytes. Thus, EBN may be suggested as a potential agent for the treatment of osteoarthritis.

7.6 Eye Care Properties

Eyes are the sensory organs that react to light resulting into the sensation of sight. Rod and cone cells forming retina allow conscious light perception and vision. Human eye distinguishes approximately 10 million colours [38]. Cornea forms the transparent frontal portion of the eye covering iris, pupil and the anterior chamber; and consists of three cell layers *viz.* epithelium, stroma and endothelium. Each cell layer carries specific function and ensure the optimal functioning of cornea in normal vision in addition to acting as protective barrier from external environment [39]. Corneal stroma, which is filled by keratocytes bound by extracellular matrix forms about 90% of the corneal volume [40,41].

Keratocytes originate from mesenchyma of the corneal stroma and ensure the synthesis and maintenance of the extracellular matrix (ECM) components [42]. The cornea is generally damaged by injuries such as abrasions, localized burns or surface or depth injuries [43]. For the development of medications for the care of keratocytes; Zainal Abidin et al. [44] demonstrated the effects of EBN on rabbit corneal keratocytes. The EBN effects were investigated on six New Zealand White Rabbits using MTT assay in FDS and FD. The highest cell proliferation was observed when both media were supplied with 0.05% and 0.1% EBN, and cell proliferation was consistently higher in FDS compared to FD. The corneal keratocytes conserved their phenotypes with EBN addition, which was confirmed by phase contrast micrographs and gene expression analysis. This report revealed the fact that low EBN concentration synergistically induced cell proliferation, especially in serum-containing medium. This is a very important breakthrough since both cellular proliferation and proper functioning maintenance are essential during corneal wound healing.

7.7 Neuroprotective Properties

Neurodegeneration involves the progressive loss of the structure and function of the basic units of nervous system, neurons. Neurodegenerative diseases including Alzheimer's, Parkinson's and Huntington's occur as eventual results of neurodegenerative processes. Globally Parkinson's disease, an age-related progressive neurodegeneration was estimated to be prevalent in approximately 9 million people over the age of 50 years by the end of 2030 [45]. Parkinson's disease is characterized by the loss of dopaminergic neurons in substantia nigra and consequently leading to dopamine depletion in the striatum [46]. In addition, abnormal accumulation of α-synuclein has also been reported in surviving neurons [47]. The dopamine depletion wanes motor functions and causes the patients to show clinical signs including tremor, rigidity and slow responsiveness [48]. Yew et al. [49] investigated the effect of EBNEs on SH-SY5Y human neuroblastoma cells. It was observed that the crude EBNE did not cause the death of SH-SY5Y cells up to 75 μg/ml concentrations. Besides, the maximum non-toxic dose (MNTD) of water extract of EBN was double to that of crude extract. Moreover, the intensity of 6-hydroxydopamine-induced apoptotic changes in SH-SY5Y cells reduced by EBN treatment, which was clear from morphological and nuclear staining observations. Further, it was interesting to note the improvement in cell viability with crude EBN extract in comparison to the water extract. However, water extract was more potent in improving ROS build up, early apoptotic membrane phosphatidylserine externalization and the inhibition of caspase-3 cleavage. It is quite evident from this research article that EBNEs induce neuroprotective effects against 6-6-hydroxydopamine-induced degeneration of dopaminergic neurons *via* the inhibition of apoptosis and, hence, may serve as a possible nutraceutical option for the protection against oxidative stress-related neurodegenerative disorders.

7.8 Anti-oxidant Properties

There are several anti-oxidant systems within the human body that negotiate the oxidative stress from regular metabolic processes. Additionally, the dietary anti-oxidants also fight with the cell-damaging effects of free radicals. Dietary anti-oxidants may act either independently or in association with the endogenous systems, and have always been beneficent to human health. Their absence in diet causes several diseases due to unrestricted oxidative stress. Several fruits and vegetables have exhibited protective effects against some cancers and other diseases. This is the reason that the people regularly consuming anti-oxidant rich fruits and vegetables have lesser frequencies of these diseases [50].

EBN has been shown to display anti-oxidant properties as it contains several bioactive compounds *viz.* glucosamine, lactoferrin, sialic acid, amino acids, fatty acids, triacylglycerol, minerals, vitamins and other anti-oxidants [51,52]. The *in vitro* bioaccessibility and anti-oxidant properties of water extracts of EBN were documented by Yida et al. [53] by using ABTS and oxygen radical absorbance capacity (ORAC) assays. The undigested water extract of EBN demonstrated a little anti-oxidant activity (1 and 1%, respectively at 1000 μg/mL) in comparison to the digested samples at similar concentrations (38 and 50%, respectively). Importantly, the EBN extracts were non-toxic towards HEPG2 cells and showed protective effects from hydrogen peroxide induced-toxicity towards HEPG2 cells. This study indicated that the digestion in gut releases the bioactive components of EBN from their matrix, which are then absorbed by passive transport. However, *in vivo* studies are needed to determine their further clinical significance.

7.9 Miscellaneous Properties

Obviously, EBN is a nutritious food material and displays a broad spectrum of biological activities. The presence of health promoting ingredients in EBN, and its use in TCM; have made it a subject of great interest. In addition to the biological activities of EBN mentioned in the above sub-sections, it has exhibited several other biologically important properties such as lectin binding properties, cure to asthma, dry coughs, tuberculosis, stomach ulcer and gastric troubles.

There has been a great demand of skin care therapies including creams, analgesics, anti-acne and moisturizers throughout the world. EBN has the reputation for skin texture and complexion improving properties among the Chinese communities. Its consumption is known to help retain youthfulness and a clean and clear facial complexion. EBN also finds occasional uses for skin texture improvement in babies with frequent skin rash outbreaks [54]. Zhang [55] developed a skin-whitening/moisturizing eye mask with compositions of sargassum (5-10), EBN (15-20), hydrolyzed pearl (5-10), aloe juice (5-10), angelica dahurica powder (3-5), soybean protein (5-10), euphorbia lathyris (3-5), butanediol (5-8), tri-thylglycine (5-10) by weight parts. The formulation was very useful for moisturizing and nourishing skin around eyes. Besides, it was also quite useful for the treatment of dark circles around eyes, and the repair of the elasticity of skin around eyes. Li and Peng [56] developed a super elastic moisturizing face mask essence with compositions of WSK Tremella fuciformis extract (0.05-0.50), sodium hyaluronate (0.01-0.20), Apus affinis Nidus *Collocaliae* (EBN) extract (0.5-2.0), hydrolyzed soy protein (0.5-2.0), allantoin (0.1-1.0), EDTA-disodium (0.01-0.2), glycerol (2.0-10.0), betaine (1.0-10.0), propanediol (1.0-10.0), carbopol (0.1-0.5), triethanolamine (0.1-0.5), flavour and antiseptic (0.3-1.0) by weights percents, and water as balance. Tremella fuciformis extract and hyaluronic acid continuously ensured the release of moisturizing factor, and lock moisture and vitamin E. This mechanism leads to skin nourishment and moisturization, and thereby increasing skin activity along with the enhancement of skin barrier to obtain ideal moisture state.

EBN is very useful to people with poor digestion, recovering from illness, and children with weak appetite. It is an easily digestible food and delivers essential nutrients into the body and therefore, brings health to those who are weak. [54]. Good quality EBN consumption is beneficent for the restoration of lung functions and in people with dry cough and periodic sputum stains, e.g., in heavy smokers [54]. EBN is very beneficent to pregnant women before and after child birth. Pregnant mothers who consume EBN recover faster after delivery of healthy and fair skinned babies, and experience considerably lesser hair loss. For post-pregnancy health of women, the supplementary EBN drinks provide more energy, better sleep, and give the feeling of vitality to the mothers [54].

Nephritis is a painful disorder characterized by the inflammation of kidneys. Glomeruli, tubules and even interstitial tissues surrounding the glomeruli and tubules are jointly involved in this complication [57]. Nephritis is basically caused by infections, some toxins and auto-immune disorders. It is a serious medical condition and represents the eighth highest cause of death in human beings. Xu [58] prepared a Chinese medical formulation with EBN as one of the ingredients for the successful therapeutic treatment of nephritis. The formulation was prepared by slicing Panax quinquefolium (3-5), soaking EBN (3-5) with water, adding chicken soup (80-85) and smashing rock candy (8-10) in percent masses. Finally, all the contents were decocted for 2 h. The preparation was highly useful in providing therapeutic relief to nephritic patients.

In conclusion, it can be said that EBN is a versatile food with a range of beneficial health effects. Several cosmetical products using EBN as one of the ingredients have already hit the market with great responses. However, its investigations for the treatment of diseases are in early stages. Therefore, more research is needed to fully explore its potential as a product of breakthrough for its medicinal properties.

Some of the famous cosmetical products having EBN as the sole or one of the ingredients are shown in **Fig. 7.1: (a-i)**.

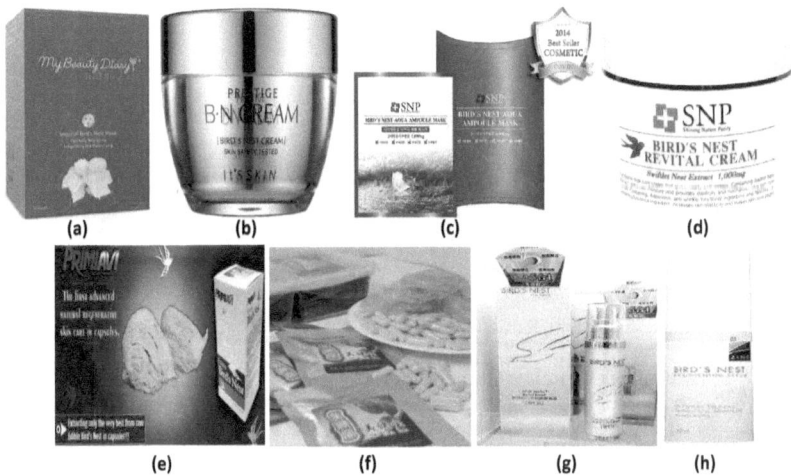

Fig. 7.1: Market looks of some of the famous cosmetical products based on EBN: **(a):** Imperial bird's nest mask, **(b):** Prestige bird's nest cream, **(c):** Bird's nest facial mask, **(d):** Bird's nest revital cream, **(e) and (f):** Natural regenerative skin care capsules, **(g):** Watson's bird's nest white perfect facial lotion and **(h):** Bird's nest brightening serum.

References

1. Ali, I., Wani, W. A., & Saleem, K. (2011). Cancer scenario in India with future perspectives. Cancer Therapy, 8, 56-70.

2. Ali, I., Wani, W. A., Haque, A., & Saleem, K. (2013). Glutamic acid and its derivatives: candidates for rational design of anticancer drugs. Future Medicinal Chemistry, 5, 961-978.

3. Ali, I., Rahis-Uddin, Salim, K., Rather, M. A., Wani, W. A., & Haque, A. (2011). Advances in nano drugs for cancer chemotherapy. Current Cancer Drug Targets, 11, 135-146.

4. Ali, I., Wani, W. A., Saleem, K., & Hsieh, M. F. (2014). Anticancer metallodrugs of glutamic acid sulphonamides: in silico, DNA binding, hemolysis and anticancer studies. RSC Advances, 4, 29629-29641.

5. Ali, I., Wani, W. A., Saleem, K., & Hsieh, M. F. (2013). Design and synthesis of thalidomide based dithiocarbamate Cu(II), Ni(II) and Ru(III) complexes as anticancer agents. Polyhedron, 56, 134-143.

6. Rashed, A. A., & Wan Nazaimoon, W. M. (2010). Effect of Edible Bird's Nest on Caco-2 Cell Proliferation. Journal of Food Technology, 8, 126-130.

7. National Center for Complementary and Alternative Medicine (NCCAM). The Use of Complementary Medicine in the United States. http://nccam.nih.gov/news/camsurvey_fsl.htm.

8. Lim, J., Wong, M., Chan, M. Y., Tan, A. M., Rajalingam, V., Lim, L. P., Lou, J., & Tan, C. L. (2006). Use of complementary and alternative medicine in paediatric oncology patients in Singapore. Annals Academy of Medicine Singapore, 35, 753-758.

9. Shih, V., Chiang, J. Y., Chan, A. (2009). Complementary and alternative medicine (CAM) usage in Singaporean adult cancer patients. Annals of Oncology, 20, 752-757.

10. Koonin, E. V., Senkevich, T. G., & Dolja, V. V. (2006). The ancient Virus World and evolution of cells. Biology Direct, 1, 29.

11. Roulston, A., Marcellus, R. C., & Branton, P. E. (1999). Viruses and apoptosis. Annuual Review of Microbiology, 53, 577-628.

12. Wang, C. C. (1921). The composition of Chinese edible birds' nests and the nature of their proteins. Journal of Biological Chemistry, 49, 429-439.

13. Guo, C. T., Takahashi, T., Bukawa, W., Takahashi, N., Yagi, H., Kato, K., Hidari, K. I., Miyamoto, D., Suzuki, T., & Suzuki, Y. (2006). Edible bird's nest inhibits influenza virus infection. Antiviral Research, 70, 140-146.

14. Howe, C., Lee, L. T., & Rose, H. M. (1961). Collocalia mucoid: A substrate for myxovirus neuraminidase. Archives of Biochemistry and Biophysics, 95, 512-520.

15. Howe, C., Lee, L. T., & Rose, H. M. (1961). Influenza virus sialidase. Nature, 188, 251-252.

16. Yagi, H., Yasukawa, N., Yu, S. Y., Guo, C. T., Takahashi, N., Takahashi, T., Bukawa, W., Suzuki, T., Khoo, K. H., Suzuki, Y., Kato, K. (2008). The expression of sialylated high-antennary N-glycans in edible bird's nest. Carbohydrate Research, 343, 1373-1377.

17. Zuk, P. A., Zhu, M., Mizuno, H., Huang, J., Futrell, J. W., Katz, A. J., Benhaim, P., Lorenz, H. P., Hedrick, M. H. (2001). Multilineage cells from human adipose tissue: implications for cell-based therapies. Tissue Engineering, 7, 211-28.

18. Gronthos, S., Franklin, D. M., Leddy, H. A., Robey, P. G., Storms, R. W., & Gimble, J. M. (2001). Surface protein characterization of human adipose tissue-derived stromal cells. Journal of Cellular Physiology, 189, 54-63.

19. Zuk, P. A., Zhu, M., Ashjian, P., De Ugarte, D. A., Huang, J. I., Mizuno, H., Alfonso, Z. C., Fraser, J. K., Benhaim, P., & Hedrick, M. H. (2002). Human adipose tissue is a source of multipotent stem cells. Molecular Biology of the Cell, 13, 4279-4295.

20. Boone, C., Mourot, J., Gregoire, F., & Remacle, C. (2000). The adipose conversion process: regulation by extracellular and intracellular factors. Reproduction Nutrition Development, 40, 325-358.

21. Wickham, M. Q., Erickson, G. R., Gimble, J. M., Vail, T. P., & Guilak, F. (2003). Multipotent stromal cells derived from the infrapatellar fat pad of the knee. CORR- Clinical Orthopaedics and Related Research, 412, 196-212.

22. Lee, J. A., Parrett, B. M., Conejero, J. A., Laser, J., Chen, J,, Kogon, A. J., Nanda, D., Grant, R. T., & Breitbart, A. S. (2003). Biological alchemy: engineering bone and fat from fat-derived stem cells. Annals of Plastic Surgery, 50, 610-617.

23. Gimble, J., & Guilak, F. (2003). Adipose-derived adult stem cells: isolation, characterization, and differentiation potential. Cytotherapy, 5, 362-369.

24. Gimble, J. M., & Guilak, F. (2003). Differentiation potential of adipose derived adult stem (ADAS) cells. Current Topics in Developmental Biology, 58, 137-160.

25. Tholpady, S. S., Llull, R., Ogle, R. C., Rubin, J. P., Futrell, J. W., & Katz, A. J. (2006). Adipose tissue: stem cells and beyond. Clinics in Plastic Surgery, 33, 55-62.

26. Aust, L., Devlin, B., Foster, S. J. et al. (2004). Yield of human adipose-derived adult stem cells from liposuction aspirates. Cytotherapy, 6, 7-14.

27. Safford, K. M., & Rice, H. E. (2005). Stem cell therapy for neurologic disorders: therapeutic potential of adipose-derived stem cells. Current Drug Targets, 6, 57-62.

28. Gimble, J. M. (2003). Adipose tissue-derived therapeutics. Expert Opinions on Biological Therapy, 3, 705-713.

29. Ogawa, R. (2006). The importance of adipose-derived stem cells and vascularized tissue regeneration in the field of tissue transplantation. Current Stem Cell Research and Therapy, 1, 13-20.

30. Roh, K. B., Lee, J., Kim, Y. S., Park, J., Kim, J. H., Lee, J., & Park, D. Mechanisms of Edible Bird's Nest Extract-Induced Proliferation of Human Adipose-Derived Stem Cells. Evidence-Based Complementary and Alternative Medicine, 2012, Article ID 797520.

31. Harris, R. C., Chung, E., & Coffey, R. J. (2003). "EGF receptor ligands". Experimental Cell Research, 284, 2-13.

32. Carpenter, G., & Cohen, S. (1990). "Epidermal growth factor". The Journal of Biological Chemistry, 265, 7709-7712.

33. Dawson, J. P., Berger, M. B., Lin, C. C., Schlessinger, J., Lemmon, M. A., & Ferguson, K. M. (2005)."Epidermal growth factor receptor dimerization and activation require ligand-induced conformational changes in the dimer interface". Molecular and Cellular Biology, 25, 7734-7742.

34. Fallon, J. H., Seroogy, K. B., Loughlin, S. E., Morrison, R. S., Bradshaw, R. A., Knaver, D. J., & Cunningham, D. D. (1984). "Epidermal growth factor immunoreactive material in the central nervous system: location and development". Science, 224, 1107-9.

35. Lin, J. R., Zhou, H., Lai, X. P., Hou, Y., Xian, X. M., Chen, J. N., Wang. P. X., Zhou, L., Dong, Y. (2009). Genetic identification of edible bird's nest based on mitochondrial DNA sequences. Food Research International, 42, 1053-1061.

36. Matsukawa, N., Matsumoto, M., Bukawa, W., Chiji, H., Nakayama, K., Hara, H., & Tsukahara, T. (2011). Improvement of bone strength and dermal thickness due to dietary edible bird's nest extract in ovariectomized rats. Bioscience Biotechnology and Biochemistry, 75, 590-592.

37. Chua, K. H., Lee, T. H., Nagandran, K., Md. Yahaya, N. H., Lee, C. T., Tjih, E. T., & Abdul Aziz, R. (2013). Edible Bird's nest extract as a chondro-protective agent for human chondrocytes isolated from osteoarthritic knee: in vitro study. BMC Complementary and Alternative Medicine, 13, 19.

38. Land, M. F., & Fernald, R. D. (1992). The evolution of eyes". Annual Review of Neuroscience, 15, 1-29.

39. Lu, L., Reinach, P. S., & Kao, W. W. Y. (2001). Corneal epithelial wound healing. Experimental Biology and Medicine, 226, 653-664.

40. Krachmer, J. H., Mannis, M. J., & Holland, E. J. (2004). Cornea: Fundamentals of cornea and external disease. Mosby-Year Book Publication, p.p.183-195.

41. West-Mays, J. A., & Dwivedi, D. J. (2006). The keratocyte: corneal stromal cell with variable repair phenotypes. International Journal of Biochemistry & Cell Biology, 38, 1625-1631.

42. He, J., & Bazan, H. E. P. (2008). Epidermal growth factor synergism with TGF-β1 via PI-3 Kinase activity in corneal keratocyte differentiation. Invest Opthalmol Vis Sci, 49, 2936-2945.

43. Gnanadoss, A. S. (2008). Manual of cornea. Jaypee Bros. Medical Publishes (P) Ltd., p.p. 62-64.

44. Zainal Abidin, F., Hui, C. K., Luan, N. S., Mohd Ramli, E. S., Hun, L. T., & Abd Ghafar, N. (2011). Effects of edible bird's nest (EBN) on cultured rabbit corneal keratocytes. BMC Complementary and Alternative Medicine, 11, 94.

45. Dorsey, E. R., Constantinescu, R., Thompson, J. P., Biglan, K. M., Holloway, R. G., Kieburtz, K., Marshall, F. J., Ravina, B. M., Schifitto, G., Siderowf, A., & Tanner, C. M. (2005). Projected number of people with Parkinson disease in the most populous nations, 2005 through 2030. Neurology, 68, 384-386.

46. Dauer, W., & Przedborski, S. (2003). Parkinson's disease: mechanisms and models. Neuron, 39, 889-909.

47. Przedborski, S. (2005). Pathogenesis of nigral cell death in Parkinson's disease. Parkinsonism and Related Disorders, 11, S3-S7.

48. Snyder, C. H., & Adler, C. H. (2007). The patient with Parkinson's disease: part I-treating the motor symptoms; part II-treating the nonmotor symptoms. Journal of the American Academy of Nurse Practitioners, 19, 179-197.

49. Yew, M. Y., Koh, R. Y., Chye, S. M., Othman, I., & Ng, K. Y. (2014). Edible bird's nest ameliorates oxidative stress-induced apoptosis in SH-SY5Y human neuroblastoma cells. BMC Complementary and Alternative Medicine, 14, 391.

50. http://www.news-medical.net/health/Antioxidant-Health-Effects.aspx. Last accessed on 07-10-2015.

51. Hamzah, Z., Ibrahim, N. H., Jaafar, M. N., Lee, B. B., Hashim, O., & Hussin, K. (2013). Nutritional properties of edible bird nest. Journal of Asian Scientific Research, 3, 600-607.

52. Liu, X., Lai, X., Zhang, S., Huang, X., Lan, Q., Li, Y., Li, B., Chen, W., Zhang, Q., Hong, D., & Yang, G. (2012). Proteomic profile of edible bird's nest proteins. Journal of Agricultural and Food Chemistry, 60, 12477-12481.

53. Yida, Z., Imam, M. U., & Ismail, M. (2014). In vitro bioaccessibility and antioxidant properties of edible bird's nest following simulated human gastro-intestinal digestion. BMC Complementary and Alternative Medicine, 14, 468.

54. http://kangshengagri.com/knowledge-hub/major-health-benefits-of-edible-bird-nests/?&lang=en. Accessed on 22-01-2015.

55. Zhang, Q. (2010). Skin-whitening/moisturizing eye mask containing extracts from Chinese medicines. Patent No.: CN 101756879.

56. Li, D., & Peng, Z. (2013). Super elastic moisturizing face mask essence and preparation method thereof. Patent No.: CN 1030554771.

57. Gardner Jr, K. D. (1971). "Athletic nephritis: Pseudo and real". Annals of internal medicine 75, 966-967.

58. Xu, F. (2010). Chinese medical composition containing Panax and Edible bird's nest for treating nephritis. Medical preparations for treating nephritis, with definite and stable therapeutic effect. Patent No.: CN 101683370.

Chapter-8: Conclusions and Future Directions of Research

EBN is a source of vital health promoting constituents including proteins, carbohydrates, fatty acids, amino acids and minerals. It has served as a food and tonic for ages, and is quite popular in the Southeast Asian region of the world. Owing to food and medicinal values, EBN has achieved the status of a highly prestigious bioproduct. Different kinds of food products *viz.* energy drinks, food additives, instant foods, etc. have been prepared from EBN. Lately, EBN and its extracts have been used in several pharmaceutical preparations mostly in cosmetics. Several studies have documented the potential of EBN and its extracts for eye care, bone strengthening, cell proliferating, anti-oxidant and antiviral properties. EBN has formed a lucrative industry in Southeast Asian region and the people are getting huge money from EBN harvesting. Swiftlets have been a constant source of income in these regions. Bird houses are being built to attract swiftlets so that nests can be harvested at safe places without risky and painful nest collection procedures. EBN is a highly cared bioproduct and any disturbance to its precursors (swiftlets) or the product itself is not acceptable.

Despite of many studies being carried out on EBN, a number of challenges are at the forefront of scientific communities to fully explore this incredible salivary bioproduct. It is the duty of scientists and other workers in this field to sort out the problems, smoothen the life of swiftlets, carry out extensive and exhaustive investigations so that more and more can be known about EBN, and its health and food effects. Besides, there is need to make EBN visible to the whole world as several populations in the world are unaware of its mysterious health and food properties. This may lead to benefit of both consumers and harvesters.

The high cost and increasing demands of EBN have led to its adulteration. Presently, several agents (karaya gum, tremella fungus, red seaweed, gelatin, fried pork skin, egg white, soybean, milk, rice, starch, jelly, agar, and fish vesicae etc.) are being adulterated into EBN to raise its weight for extra benefits. Besides, some other adulteration procedures such as bleaching, staining, and incorporating cheaper and low quality EBN materials into the more expensive ones; are being frequently used. More importantly, some harmful substances have been detected in EBN. Nitrites and *staphylococcus aureus* have been detected in imported EBN samples in China [1]. Additionally, semicarbazide, a known potential harmful chemical agent was found in EBN. Xin et al. [2] detected semicarbazide in instant bottled EBN samples and claimed that semicarbazide might have originated from the bleaching process used in EBN preparation. These adulterants and the chemical agents are very harmful to the human beings who consume EBN. Not only this, these things can also defame EBN as food and medicine. Therefore, for ensuring the quality of EBN, both efficient animal husbandry practice and proper EBN handling are crucial steps to be taken and timely improved. It is important that swiftlet ranches strictly comply with the requirements laid down by the authorities, and adequate enough cleaning is applied to raw unclean EBN. As discussed in the previous chapters, scientifically trained workers are needed for the cleaning and processing of EBN. Besides, care is to be taken for the standard of agents that are worked with during cleaning processes. Moreover, authentication procedures of EBN have to be at their best in order to prevent EBN adulteration with fake materials. All the EBN samples should be subjected to authentication tests before sale.

Despite of the limited science available on the biological and health-promoting effects of EBN in the past, several scientific publications and patents have appeared lately. Some of the publications have revealed the immune-enhancing, bone-strengthening, antiviral, anticancer, cell proliferation enhancing and anti-skin aging effects of EBN. However, more scientific research is required before concluding on the health-promoting effects of EBN. There is need to fully explore the mechanistic basis as how does the consumption of EBN alleviate asthma, improve concentration, facilitate renal function, improve complexion, stamina and vitality. It will be good to know which fraction of EBN is active and which is not. The specific components leading to specific functions are needed to be identified. Importantly, the correlations between dosages and activities of EBN need to be worked out. Therefore, the revelation of the underlying mechanisms by which EBN exerts both its *in vitro* and *in vivo* therapeutic effects would be a great breakthrough. This would be even bettered if the use of EBN is followed up in a clinical setting to investigate its health promoting and therapeutic values.

Some metabolite profiling studies of EBN have been documented, however, there are no rational correlations that could relate the specific effects of EBN to its different components. Besides, the components contributing to the biological effects of EBN have not been isolated and purified yet. Therefore, the present era demands more proteomic and genomic studies for the thorough investigation of EBN and its components for the welfare of mankind. The literature updates indicate that EBNs obtained from different sources and locations have differences in composition and the biological effects thereof. Therefore, it would be really of benefit, if standardization of EBN formulations can be achieved so that a more predictable outcome can be obtained for using EBN in various conditions.

Toxicity and toxicity management issues are of utmost consideration for the use and consumption of EBN as food and medicine. EBN was proved as the most common cause of food-induced anaphylaxis among children in some clinical tests in the National University of Singapore. The experiments suggested that immunoglobulin E-mediated process was involved in anaphylaxis, and five major putative allergens were characterized [3]. A similar case of EBN related toxicity was reported in Japan. A Malaysian student developed erythema with facial swelling and nasal obstruction within five minutes of consumption of a dessert containing EBN. Such allergic reactions are sometimes life-threatening, and the identification of the allergen is the most effective means of management [3]. Therefore, it is very crucial to know whether a person is allergic towards some EBN protein or some other ingredient, which will cause anaphylaxis. Hence, skin prick tests are encouraged before consumption by first time consumers.

There are some important issues in relation to the consumption of EBN by cancer patients. Basically, EGF receptors are highly expressed in several solid tumours including breast, head and neck, non-small-cell lung, ovarian and colon, and renal cancers [4]. Thus, it may be envisaged that EBN consumption in such cancer patients may stimulate tumour progression and exhibit resistance to chemotherapy or radiation treatment. Therefore, it looks better for cancer patients to avoid EBN consumption.

Overall, EBN is a prized bioproduct for the people in Southeast Asian region. EBN industry has become a source of income for a large group of people. Therefore, it becomes necessary to take care of the swiftlets that are gifting this prized bioproduct. Besides, there is a need to bring the food and medicinal values of EBN into the notice of people all over the world. Obviously, it would be better if people other than those living in Southeast Asia enjoy the health benefits of EBN. Hopefully, all the measures discussed in this textbook will help in providing neat and pure nests to the mankind in need.

References

1. Zuo, Y. Nitrite found in imported edible bird's nests. People's Daily. Available at: http://english.peopledaily.com. cn/90882/7689478.html. Last accessed on 07-10-2015.

2. Xin, Y. N., Ni, H. G., & Chen, Z. Y. (2012). Semicarbazide in selected bird's nest products. Journal of Food Protection, 75, 1645-1649.

3. Denise, L. M., Goh, F. T. C., Chua, K. Y., Chay, O. M., & Lee, B. W. (20002). Edible "bird's nest"-induced anaphylaxis: An under-recognized entity? The Journal of Pediatrics, 137 (2), 277-279.

4. Herbst, R. S., & Shin, D. M. (2002). Monoclonal Antibodies to Target Epidermal Growth Factor Receptor-Positive Tumors. Cancer, 94, 1593-1611.

Abbreviations

^{13}C NMR: Carbon-13 nuclear magnetic resonance
^{1}H NMR: Proton nuclear magnetic resonance
2-DE: Two-dimensional gel electrophoresis
3T3 fibroblasts: Primary mouse embryonic fibroblasts
ABTS: 2,2-azinobis-(3-ethylbenzothiazoline-6-sulfonic acid)
AP-1: Activator protein 1
ASCs: Adipose stem cells
bp: Base pairs
Caco-2 cells: Human epithelial colorectal adenocarcinoma cells
CAM: Complementary and alternative medicine
COX-2: Cyclooxygenase-2
DNA: Deoxyribonucleic acid
EBN: Edible bird's nest
EBNE: Edible bird's nest extract
ECM: Extracellular matrix
EDTA-disodium: Disodium salt of ethylenediaminetetraacetate
EGF: Epidermal growth factor
EGFR: Epidermal growth factor receptor
FD: serum-free medium
FDS: serum-containing medium
GC: Gas chromatography
HACs: Human articular chondrocytes
hADSCs: Human adipose-derived stem cells
HEPG2 cells: Human hepatocellular carcinoma
IEC: Ion-exchange chromatography
IgE: Immunoglobulin E
IL-1: Interleukin-1
IL-6: Interleukin-6
IL-8: Interleukin-8
iNOS: Inducible nitric oxide synthase
JNK: c-Jun N-terminal kinase
kDa: Kilo Dalton
MALDI-TOF-MS: Matrix assisted laser desorption/ionization-time of flight-mass spectrometry
MDCK cells: Madin-Darby canine kidney epithelial cells
MMP1: Matrix metalloproteinase-1
MMP3: Matrix metalloproteinase-3
MNTD: Maximum non-toxic dose
MS: Mass spectrometry
mt-DNA: Mitochondrial deoxyribonucleic acid

MTT: 3-(4,5-dimethylthiazol-2-yl)-2,5-diphenyltetrazolium bromide
NADH: Nicotinamide adenine dinucleotide hydrogen
NCBI: National centre for biotechnology information
NF-κB: Nuclear factor kappa-light-chain-enhancer of activated B cells
OA: Osteoarthritis
ORAC: Oxygen radical absorbance capacity
p38 MAPK: P38 mitogen-activated protein kinase
p44/42 MAPK: p44/42 mitogen-activated protein kinase
PDTC: Ammonium pyrrolidinedithiocarbamate
PGE2: Prostaglandin E2
ppm: Parts per million
RM: Ringgit Malaysia
ROS: Reactive oxygen species
rRNA: Ribosomal ribonucleic acid
SDS: Sodium dodecyl sulphate
sGAG: Total sulphated glycosaminoglycan
SH-SY5Y: Human neuroblastoma cells
SOX-9: SRY (sex determining region Y)-box 9
TCM: Traditional Chinese Medicine
US FDA: United States Food and Drug Administration
VEGF: Vascular endothelial growth factor

I want morebooks!

Buy your books fast and straightforward online - at one of the world's fastest growing online book stores! Environmentally sound due to Print-on-Demand technologies.

Buy your books online at
www.get-morebooks.com

Kaufen Sie Ihre Bücher schnell und unkompliziert online – auf einer der am schnellsten wachsenden Buchhandelsplattformen weltweit!
Dank Print-On-Demand umwelt- und ressourcenschonend produziert.

Bücher schneller online kaufen
www.morebooks.de

OmniScriptum Marketing DEU GmbH
Heinrich-Böcking-Str. 6-8
D - 66121 Saarbrücken
Telefax: +49 681 93 81 567-9

info@omniscriptum.com
www.omniscriptum.com

www.ingramcontent.com/pod-product-compliance
Lightning Source LLC
Chambersburg PA
CBHW020449220526

45464CB00002B/927